Alan Chalmers is Adjunct Associate Professor at the University of Sydney, where he taught from 1971, first in the School of Philosophy, and from 1987 in the Unit for the History and Philosophy of Science, which he was instrumental in setting up. Born in Bristol, England, in 1939, he graduated in physics at the University of Bristol in 1961, and received an MSc in physics from the University of Manchester in 1964. He taught physics and the history of science for two years before returning to full-time study at the University of London, where he received his PhD in history and philosophy of science in 1971. He was elected a Fellow of the Academy of Humanities in 1997 and was a Visiting Scholar in the Philosophy Department at Flinders University from 2000 to 2010.

FOURTH EDITION

WHAT IS THIS THING CALLED SCIENCE?

ALAN CHALMERS

UQP

First published 1976 by University of Queensland Press
PO Box 6042, St Lucia, Queensland 4067 Australia
Second edition 1982
Third edition 1999
Reprinted 2000, 2002, 2003, 2004, 2006, 2007, 2010
Fourth edition 2013
Reprinted 2017, 2018, 2020, 2021, 2023

www.uqp.com.au
uqp@uqp.uq.edu.au

Typeset in 10.5/12.5 pt Bembo by Post Pre-press Group, Brisbane
Printed in Australia by McPherson's Printing Group

Cataloguing-in-Publication entry is available from the
National Library of Australia
http://catalogue.nla.gov.au/

Chalmers, A.F. (Alan Francis), 1939– .
What is this thing called science? : an assessment of the nature and status of science and its methods.
1. Science – Philosophy. I. Title.
4th ed.
Bibliography.
Includes index.
501
ISBN 978 0 7022 4963 1

MIX
Paper | Supporting responsible forestry
FSC® C001695

'Like all young men I set out to be a
genius, but mercifully laughter intervened.'

Clea Lawrence Durrell

Contents

Preface to the first edition

This book is intended to be a simple, clear and elementary introduction to modern views about the nature of science. When teaching philosophy of science, either to philosophy undergraduates or to scientists wishing to become familiar with recent theories about science, I have become increasingly aware that there is no suitable single book, or even a small number of books, that one can recommend to the beginner. The only sources on the modern views that are available are the original ones. Many of these are too difficult for beginners, and in any case they are too numerous to be made easily available to a large number of students. This book will be no substitute for the original sources for anyone wishing to pursue the topic seriously, of course, but I hope it will provide a useful and easily accessible starting point that does not otherwise exist.

My intention of keeping the discussion simple proved to be reasonably realistic for about two-thirds of the book. By the time I had reached that stage and had begun to criticise the modern views, I found, to my surprise, first, that I disagreed with those views more than I had thought and, second, that from my criticism a fairly coherent alternative was emerging. That alternative is sketched in the latter chapters of the book. It would be pleasant for me to think that the second half of this book contains not only summaries of current views on the nature of science but also a summary of the next view.

My professional interest in history and philosophy of science began in London, in a climate that was dominated by the views of Professor Karl Popper. My debt to him, his writings, his lectures and his seminars, and also to the late Professor Imre Lakatos, must be very evident from the contents of this book. The form of the first half of it owes much to Lakatos's brilliant article on the methodology of research programs. A noteworthy feature of the Popperian school was the pressure it put on one to be clear about the problem one was interested in and to express one's views on it in a simple and straightforward way. Although I owe much to the example of Popper and Lakatos in this respect, any ability that I have to express myself simply and clearly stems mostly from my interaction with Professor Heinz Post, who was my supervisor at Chelsea College while I was working on my doctorial thesis in the Department of History and Philosophy of Science there. I cannot rid myself of an uneasy feeling that his copy of this book will be returned to me along with the demand that I rewrite the bits he does not understand. Of my colleagues in London to whom I owe a special debt, most of them students at the time, Noretta Koertge, now at Indiana University, helped me considerably.

I referred above to the Popperian school as a *school*, and yet it was not until I came to Sydney from London that I fully realised the extent to which I had been in a school. I found, to my surprise, that there were philosophers influenced by Wittgenstein or Quine or Marx who thought that Popper was quite wrong on many issues, and some who even thought that his views were positively dangerous. I think I have learnt much from that experience. One of the things that I have learnt is that on a number of major issues Popper is indeed wrong, as is argued in the latter portions of this book. However, this does not alter the fact that the Popperian approach is infinitely better than the approach adopted in most philosophy departments that I have encountered.

I owe much to my friends in Sydney who have helped to waken me from my slumber. I do not wish to imply by this that I accept their views rather than Popperian ones. They know better than that. But since I have no time for obscurantist nonsense about the incommensurability of frameworks (here Popperians prick up their ears), the extent to which I have been forced to acknowledge and counter the views of my Sydney colleagues and adversaries has led me to understand the strengths of their views and the weaknesses of my own. I hope I will not upset anyone by singling out Jean Curthoys and Wal Suchting for special mention here.

Lucky and attentive readers will detect in this book the odd metaphor stolen from Vladimir Nabokov, and will realise that I owe him some ackowledgment (or apology).

I conclude with a warm 'hello' to those friends who don't care about the book, who won't read the book, and who had to put up with me while I wrote it.

Alan Chalmers
Sydney, 1976

Preface to the second edition

Judging by responses to the first edition of this book it would seem that the first eight chapters of it function quite well as 'a simple, clear and elementary introduction to modern views about the nature of science'. It also seems to be fairly universally agreed that the last four chapters fail to do so. Consequently, in this revised and extended edition I have left chapters 1–8 virtually unchanged and have replaced the last four chapters by six entirely new ones. One of the problems with the latter part of the first edition was that it ceased to be simple and elementary. I have tried to keep my new chapters simple, although I fear I have not entirely succeeded when dealing with the difficult issues of the final two chapters. Although I have tried to keep the discussion simple, I hope I have not thereby become uncontroversial.

Another problem with the latter part of the first edition is lack of clarity. Although I remain convinced that most of what I was groping for there was on the right track, I certainly failed to express a coherent and well-argued position, as my critics have made clear. Not all of this can be blamed on Louis Althusser, whose views were very much in vogue at the time of writing, and whose influence can still be discerned to some extent in this new edition. I have learnt my lesson and in future will be very wary of being unduly influenced by the latest Paris fashions.

My friends Terry Blake and Denise Russell have convinced me that there is more of importance in the writings of Paul Feyerabend than I was previously prepared to admit. I have given him more attention in this new edition and have tried to separate the wheat from the chaff, the anti-methodism from the dadaism. I have also been obliged to separate the important sense from 'obscurantist nonsense about the incommensurability of frameworks'.

The revision of this book owes much to the criticism of numerous colleagues, reviewers and correspondents. I will not attempt to name them all, but acknowledge my debt and offer my thanks.

Since the revision of this book has resulted in a new ending, the original point of the cat on the cover has been lost. However, the cat does seem to have a considerable following, despite her lack of whiskers, so we have retained her, and merely ask readers to reinterpret her grin.

Alan Chalmers
Sydney, 1981

Preface to the third edition

This edition represents a major reworking of the previous edition, in which very few of the original chapters have emerged unscathed and many have been replaced. There are also a number of new chapters. The changes were necessary for two reasons. First, the teaching of an introductory course in the philosophy of science that I have undertaken in the twenty years since first writing this book has taught me how to do the job better. Second, there have been important developments in the philosophy of science in the last decade or two that need to be taken account of in any introductory text.

A currently influential school in the philosophy of science involves an attempt to erect an account of science on Bayes' theorem, a theorem in the probability calculus. A second trend, 'the new experimentalism', involves paying more attention than hitherto to the nature and role of experiment in science. Chapters 12 and 13, respectively, contain a description and an appraisal of these schools of thought. Recent work, especially that of Nancy Cartwright, has brought to the fore questions about the nature of laws as they figure in science, so a chapter on this topic is included in this new edition, as is a chapter that aims to keep abreast of the debate between realist and anti-realist interpretations of science.

So while not pretending that I have arrived at the definitive answer to the question that forms the title of this book, I have endeavoured to keep abreast of the contemporary debate and to introduce the reader to it in a way that is not too technical. There are suggestions for further reading at the end of each chapter, which will be a useful and up-to-date starting point for those who wish to pursue these matters in greater depth.

I will not attempt to name all the colleagues and students from whom I have learnt how to improve this book. I learnt much at an international symposium held in Sydney in June 1997, 'What Is This Thing Called Science? Twenty Years On'. I thank the sponsors of that symposium, The British Council, the University of Queensland Press, the Open University Press, Hackett Publishing

Company and Uitgeverij Boom, and those colleagues and old friends who attended and participated in the proceedings. The event did much to boost my morale and gave me the incentive to undertake the major task that was involved in rewriting the text. Much of the rewriting was done while I was a Research Fellow at the Dibner Institute for the History of Science and Technology, MIT, for which I express my appreciation. I could not have hoped for a more supportive environment, and one more conducive to some concentrated work. I thank Hasok Chang for his careful reading of the manuscript and his helpful comments.

I have lost track of what the cat is meant to be grinning about, but I seem to detect a note of continuing approval, which is reassuring.

Alan Chalmers
Cambridge, Mass., 1998

Preface to the fourth edition

Since this book first appeared in 1976 I have twice seen fit to write a new edition of it, removing passages, or even whole chapters, that I found unhelpful, wrong-headed or insufficiently clear, and adding new passages or chapters drawing on developments in the literature as well as on clarifications of my own thoughts. With such an end in view, I recently subjected the third edition to a critical reading. I did not find much with which I was dissatisfied as had been the case with my reappraisal of the first and second editions. I did, nevertheless, discern ways in which key themes in the book could be clarified and extended. The main source for this rethinking was the work that went into the writing of my book *The Scientist's Atom and the Philosopher's Stone: How Science Succeeded and Philosophy Failed to Gain Knowledge of Atoms*. The story of how scientific knowledge of atoms became possible proves to be a ready source of examples to illustrate and support my main points concerning the distinctive character of scientific as opposed to other kinds of knowledge. Accordingly, I have included a Postscript in this fourth edition that draws on this material to help clarify what this thing called science is.

My academic home for the first decade of the twenty-first century was the Philosophy Department at Flinders University in Adelaide. I thank my colleagues there, especially Rodney Allen, George Couvalis and Greg O'Hair, for helping to make that period productive. Of the many academics that have provided me with help, support and constructive criticism, Ursula Klein, Deborah Mayo, Alan Musgrave and John Norton deserve special mention. From 2003 to 2005 my work was supported by a grant from the Australian Research Council. I benefited from Research Fellowships at the University of Canterbury, New Zealand, and the University of Pittsburgh and from a semester in the Department of Philosophy at the University of Bristol. All of this support was very helpful and much appreciated. Sandra Grimes has been a constant and much appreciated and valued, if unduly acknowledged, support.

Alan Chalmers
Sydney, 2012

Introduction

Science is highly esteemed. Apparently it is a widely held belief that there is something special about science and its methods. The naming of some claim or line of reasoning or piece of research 'scientific' is done in a way that is intended to imply some kind of merit or special kind of reliability. But what, if anything, is so special about science? What is this 'scientific method' that allegedly leads to especially meritorious or reliable results? This book is an attempt to elucidate and answer questions of that kind.

There is an abundance of evidence from everyday life that science is held in high regard, in spite of some disenchantment with science because of consequences for which some hold it responsible, such as hydrogen bombs and pollution. Advertisements frequently assert that a particular product has been scientifically shown to be whiter, more potent, more sexually appealing or in some way superior to rival products. This is intended to imply that the claims are particularly well founded and perhaps beyond dispute. A recent newspaper advertisement advocating Christian Science was headed 'Science speaks and says the Christian Bible is provedly true' and went on to tell us that 'even the scientists themselves believe it these days'. Here we have a direct appeal to the authority of science and scientists. We might well ask what the basis for such authority is. The high regard for science is not restricted to everyday life and the popular media. It is evident in the scholarly and academic world too. Many areas of study are now described as sciences by their supporters, presumably in an effort to imply that the methods used are as firmly based and as potentially fruitful as in a traditional science such as physics or biology. Political science and social science are by now commonplace. Many Marxists are keen to insist that historical materialism is a science. In addition, Library Science, Administrative Science, Speech Science, Forest Science, Dairy Science, Meat and Animal Science and Mortuary Science have all made their appearance on university syllabuses.[1] The debate about the status of 'creation science' is still active. It is noteworthy in this context that participants on both sides of the debate assume that there is some special category 'science'. What they disagree about is whether creation science qualifies as a science or not.

Many in the so-called social or human sciences subscribe to a line of argument that runs roughly as follows. 'The undoubted success of physics over the last three hundred years, it is assumed, is to be attributed to the application of a special method, "the scientific method". Therefore, if the social and human sciences are to emulate the success of physics then that is to be achieved by first understanding and formulating this method and then applying it to the social and human sciences.' Two fundamental questions are raised by this line of argument, namely, 'what is this scientific method that is alleged to be the key to the success of physics?' and 'is it legitimate to transfer that method from physics and apply it elsewhere?'

All this highlights the fact that questions concerning the distinctiveness of scientific knowledge, as opposed to other kinds of knowledge, and the exact identification of the scientific method are seen as fundamentally important and consequential. As we shall see, however, answering these questions is by no means straightforward. A fair attempt to capture widespread intuitions about the answers to them is encapsulated, perhaps, in the idea that what is so special about science is that it is derived from the facts, rather than being based on personal opinion. This maybe captures the idea that, whereas personal opinions may differ over the relative merits of the novels of Charles Dickens and D. H. Lawrence, there is no room for such variation of opinions on the relative merits of Galileo's and Einstein's theories of relativity. It is the facts that are presumed to determine the superiority of Einstein's innovations over previous views on relativity, and anyone who fails to appreciate this is simply wrong.

As well shall see, the idea that the distinctive feature of scientific knowledge is that it is derived from the facts of experience can only be sanctioned in a carefully and highly qualified form, if it is to be sanctioned at all. We will encounter reasons for doubting that facts acquired by observation and experiment are as straightforward and secure as has traditionally been assumed. We will also find that a strong case can be made for the claim that scientific knowledge can neither be conclusively proved nor conclusively disproved by reference to the facts, even if the availability of those facts is assumed. Some of the arguments to support this skepticism are based on an analysis of the nature of observation and on the nature of logical reasoning and its capabilities. Others stem from a close look at the history of science and contemporary scientific practice. It has been a feature of modern developments in theories of science and scientific method that increasing attention has been paid to the history of science. One of the embarrassing results of this for many philosophers of science is that those episodes in the history of science that are commonly regarded as most characteristic of major advances, whether they be the innovations of Galileo, Newton, Darwin or Einstein, do not match what standard philosophical accounts of science say they should be like.

One reaction to the realisation that scientific theories cannot be conclusively proved or disproved and that the reconstructions of philosophers bear little resemblance to what actually goes on in science is to give up altogether the idea that science is a rational activity operating according to some special method. It is a reaction somewhat like this that led the philosopher Paul Feyerabend (1975) to write a book with the title *Against Method: Outline of an Anarchistic Theory of Knowledge.* According to the most extreme view that has been read into Feyerabend's later writings, science has no special features that render it intrinsically superior to other kinds of knowledge such as ancient myths or voodoo. A high regard for science is seen as a modern religion, playing a similar role to that played by Christianity in Europe in earlier eras. It is suggested that the choices between scientific theories boil down to choices determined by the subjective values and wishes of individuals.

Feyerabend's skepticism about attempts to rationalise science is shared by more recent authors from a sociological or so-called postmodernist perspective.

This kind of response to the difficulties with traditional accounts of science and scientific method is resisted in this book. An attempt is made to accept what is valid in the challenges by Feyerabend and many others, but yet to give an account of science that captures its distinctive and special features in a way that can answer those challenges.

CHAPTER 1

Science as knowledge derived from the facts of experience

A widely held commonsense view of science

In the Introduction I ventured the suggestion that a popular conception of the distinctive feature of scientific knowledge is captured by the slogan 'science is derived from the facts'. In the first four chapters of this book this view is subjected to a critical scrutiny. We will find that much of what is typically taken to be implied by the slogan cannot be defended. Nevertheless, we will find that the slogan is not entirely misguided and I will attempt to formulate a defensible version of it.

When it is claimed that science is special because it is based on the facts, the facts are presumed to be claims about the world that can be directly established by a careful, unprejudiced use of the senses. Science is to be based on what we can see, hear and touch rather than on personal opinions or speculative imaginings. If observation of the world is carried out in a careful, unprejudiced way then the facts established in this way will constitute a secure, objective basis for science. If, further, the reasoning that takes us from this factual basis to the laws and theories that constitute scientific knowledge is sound, then the resulting knowledge can itself be taken to be securely established and objective.

The above remarks are the bare bones of a familiar story that is reflected in a wide range of literature about science. 'Science is a structure built upon facts' writes J. J. Davies (1968, p. 8) in his book on the scientific method, a theme elaborated on by H. D. Anthony (1948, p. 145):

> It was not so much the observations and experiments which Galileo made that caused the break with tradition as his *attitude* to

> them. For him, the facts based on them were taken as facts, and not related to some preconceived idea . . . The facts of observation might, or might not, fit into an acknowledged scheme of the universe, but the important thing, in Galileo's opinion, was to accept the facts and build the theory to fit them.

Anthony here not only gives clear expression to the view that scientific knowledge is based on the facts established by observation and experiment, but also gives a historical twist to the idea, and he is by no means alone in this. An influential claim is that, as a matter of historical fact, modern science was born in the early seventeenth century when the strategy of taking the facts of observation seriously as the basis for science was first seriously adopted. It is held by those who embrace and exploit this story about the birth of science that, prior to the seventeenth century, the observable facts were not taken seriously as the foundation for knowledge. Rather, so the familiar story goes, knowledge was based largely on authority, especially the authority of the philosopher Aristotle and the authority of the Bible. It was only when this authority was challenged by an appeal to experience, by pioneers of the new science such as Galileo, that modern science became possible. The following account of the oft-told story of Galileo and the Leaning Tower of Pisa, taken from Rowbotham (1918, pp. 27–9), nicely captured the idea.

> Galileo's first trial of strength with the university professors was connected with his researches into the laws of motion as illustrated by falling bodies. It was an accepted axiom of Aristotle that the speed of falling bodies was regulated by their respective weights: thus, a stone weighing two pounds would fall twice as quick as one weighing only a single pound and so on. No one seems to have questioned the correctness of this rule, until Galileo gave it his denial. He declared that weight had nothing to do with the matter, and that . . . two bodies of unequal weight . . . would reach the ground at the same moment. As Galileo's statement was flouted by the body of professors, he determined to put it to a public test. So he invited the whole University to witness the experiment which he was about to perform from the leaning

> tower. On the morning of the day fixed, Galileo, in the presence of the assesmbled University and townsfolk, mounted to the top of the tower, carrying with him two balls, one weighing one hundred pounds and the other weighing one pound. Balancing the balls carefully on the edge of the parapet, he rolled them over together; they were seen to fall evenly, and the next instant, with a loud clang, they struck the ground together. The old tradition was false, and modern science, in the person of the young discoverer, had vindicated her position.

Two schools of thought that involve attempts to formalise what I have called a common view of science, that scientific knowledge is derived from the fact, are the empiricists and the positivists. The British empiricists of the seventeenth and eighteenth centuries, notably John Locke, George Berkeley and David Hume, held that all knowledge should be derived from ideas implanted in the mind by way of sense perception. The positivists had a somewhat broader and less psychologically orientated view of what facts amount to, but shared the view of the empiricists that knowledge should be derived from the facts of experience. The logical positivists, a school of philosophy that originated in Vienna in the 1920s, took up the positivism that had been introduced by Auguste Comte in the nineteenth century and attempted to formalise it, paying close attention to the logical form of the relationship between scientific knowledge and the facts. Empiricism and positivism share the common view that scientific knowledge should in some way be derived from the facts arrived at by observation.

There are two other rather distinct issues involved in the claim that science is derived from the facts. One concerns the nature of these 'facts' and how scientists are meant to have access to them. The second concerns how the laws and theories that constitute our knowledge are derived from the facts once they have been obtained. We will investigate these two issues in turn, devoting this and the next two chapters to a discussion of the nature of the facts on which science is alleged to be based and chapter 4 to the question of how scientific knowledge might be thought to be derived from them.

Three components of the stand on the facts assumed to be the basis of science in the common view can be distinguished. They are:

(a) Facts are directly given to careful, unprejudiced observers via the senses.
(b) Facts are prior to and independent of theory.
(c) Facts constitute a firm and reliable foundation for scientific knowledge.

As we shall see, each of these claims is faced with difficulties and, at best, can only be accepted in a highly qualified form.

Seeing is believing

Partly because the sense of sight is the sense most extensively used to observe the world, and partly for convenience, I will restrict my discussion of observation to the realm of seeing. In most cases, it will not be difficult to see how the argument presented could be re-cast so as to be applicable to the other senses. A simple account of seeing might run as follows. Humans see using their eyes. The most important components of the human eye are a lens and a retina, the latter acting as a screen on which images of objects external to the eye are formed by the lens. Rays of light from a viewed object pass from the object to the lens via the intervening medium. These rays are refracted by the material of the lens in such a way that they are brought to a focus on the retina, so forming an image of the object. Thus far, the functioning of the eye is analogous to that of a camera. A big difference is in the way the final image is recorded. Optic nerves pass from the retina to the central cortex of the brain. These carry information concerning the light striking the various regions of the retina. It is the recording of this information by the brain that constitutes the seeing of the object by the human observer. Of course, many details could be added to this simplified description, but the account offered captures the general idea.

Two points are strongly suggested by the foregoing account of observation through the sense of sight that are incorporated into the common or empiricist view of science. The first is that a human observer has more or less direct access to knowledge of some facts about the world insofar as they are recorded by the brain in the act of seeing. The second is that two normal observers viewing the same object or scene from the same place will 'see' the same thing. An identical combination of light rays will strike the eyes of each observer, will be focused on their normal retinas by their normal eye lenses and give rise to similar images. Similar information will then travel to the brain of each observer via their normal optic nerves, resulting in the two observers seeing the same thing. In subsequent sections we will see why this kind of picture is seriously misleading.

Visual experiences not determined solely by the object viewed

In its starkest form, the common view has it that facts about the external world are directly given to us through the sense of sight. All we need to do is confront the world before us and record what is there to be seen. I can establish that there is a lamp on my desk or that my pencil is yellow simply by noting what is before my eyes. Such a view can be backed up by a story about how the eye works, as we have seen. If this was all there was to it, then what is seen would be determined by the nature of what is looked at, and observers would always have the same visual experiences when confronting the same scene. However, there is plenty of evidence to indicate that this is simply not the case. Two normal observers viewing the same object from the same place under the same physical circumstances do not necessarily have identical visual experiences, even though the images on their respective retinas may be virtually identical. There is an important sense in which two observers need not 'see' the same thing. As N. R. Hanson (1958) has put it, 'there is more to seeing than meets the eyeball'. Some simple examples will illustrate the point.

Most of us, when first looking at Figure 1, see the drawing of a staircase with the upper surface of the stairs visible. But this is not the only way in which it can be seen. It can without difficulty be seen as a staircase with the under surface of the stairs visible. Further, if one looks at the picture for some time, one generally finds that what one sees changes frequently, and involuntarily, from a

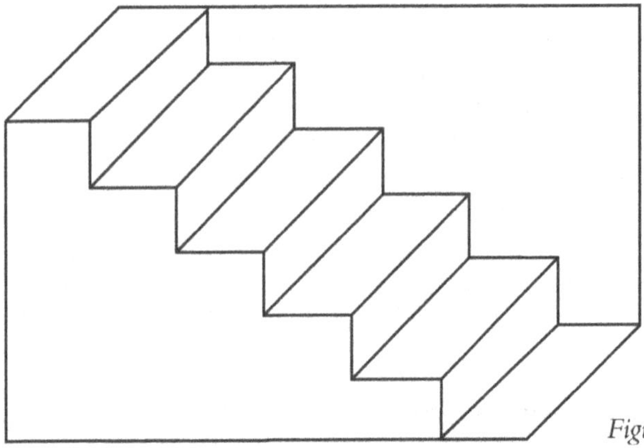

Figure 1

staircase viewed from above to one viewed from below and back again. And yet it seems reasonable to suppose that, since it remains the same object viewed by the observer, the retinal images do not change. Whether the picture is seen as a staircase viewed from above or one viewed from below seems to depend on something other than the image on the retina of the viewer. I suspect that no reader of this book has questioned my claim that Figure 1 depicts a staircase. However, the results of experiments on members of African tribes whose culture does not include the custom of depicting three-dimensional objects by two-dimensional perspective drawings, nor staircases for that matter, indicate that members of those tribes would not see Figure 1 as a staircase at all. Again, it seems to follow that the perceptual experiences that individuals have in the act of seeing are not uniquely determined by the images on their retinas. Hanson (1958, chapter 1) contains some more captivating examples that illustrate this point.

Another instance is provided by a children's picture puzzle that involves finding the drawing of a human face among the foliage in the drawing of a tree. Here, what is seen, that is, the subjective impressions experienced by a person viewing the drawing, at first corresponds to a tree, with trunk, branches and leaves. But this changes once the human face has been detected. What was once seen as branches and leaves is now seen as a human face. Again, the same physical object is viewed before and after the solution of the puzzle, and presumably the image on the observer's retina does not change at the moment the puzzle is solved and the face found. If the picture is viewed at some later time, the face is readily and quickly seen by an observer who has already solved the puzzle once. It would seem that there is a sense in which what an observer sees is affected by his or her past experience.

'What', it might well be suggested, 'have these contrived examples got to do with science?' In response, it is not difficult to produce examples from the practice of science that illustrate the same point, namely, that what observers see, the subjective experiences that they undergo, when viewing an object or scene is not determined solely by the images on their retinas but depends also on the experience, knowledge and expectations of the observer. The point is implicit in the uncontroversial realisation that one has to learn to be a competent observer in science. Anyone who has been through the experience of having to learn to see through a microscope will need no convincing of this. When the beginner looks at a slide prepared by an instructor through a microscope it is rare that the appropriate cell structures can be discerned, even though the instructor has no difficulty discerning them when looking at the same slide through the same microscope. It is significant to note, in this context, that microscopists found no great difficulty observing cells divide in suitably prepared circumstances once they were alert for what to look for, whereas prior to this discovery these cell divisions went unobserved, although we now know they must have been there to be observed in many of the samples examined through a microscope. Michael Polanyi (1973, p. 101) describes the changes in a medical student's perceptual experience when he is taught to make a diagnosis by inspecting an X-ray picture.

> Think of a medical student attending a course in the X-ray diagnosis of pulmonary diseases. He watches, in a darkened room, shadowy traces on a fluorescent screen placed against a patient's chest, and hears the radiologist commenting to his assistants, in technical language, on the significant features of these shadows. At first, the student is completely puzzled. For he can see in the X-ray picture of a chest only the shadows of the heart and ribs,

> with a few spidery blotches between them. The experts seem to be romancing about figments of their imagination; he can see nothing that they are talking about. Then, as he goes on listening for a few weeks, looking carefully at ever-new pictures of different cases, a tentative understanding will dawn on him; he will gradually forget about the ribs and begin to see the lungs. And eventually, if he perseveres intelligently, a rich panorama of significant details will be revealed to him; of physiological variations and pathological changes, of scars, of chronic infections and signs of acute disease. He has entered a new world. He still sees only a fraction of what the experts can see, but the pictures are definitely making sense now and so do most of the comments made on them.

The experienced and skilled observer does not have perceptual experiences identical to those of the untrained novice when the two confront the same situation. This clashes with a literal understanding of the claim that perceptions are given in a straightforward way via the senses.

A common response to the claim that I am making about observation, supported by the kinds of examples I have utilised, is that observers viewing the same scene from the same place see the same thing but interpret what they see differently. I wish to dispute this. As far as perception is concerned, the only things with which an observer has direct and immediate contact are his or her experiences. These experiences are not uniquely given and unchanging but vary with the knowledge and expectations possessed by the observer. What is uniquely given by the physical situation, I am prepared to admit, is the image on the retina of an observer, but an observer does not have direct perceptual contact with that image. When defenders of the common view assume that there is something unique given to us in perception that can be interpreted in various ways, they are assuming without argument, and in spite of much evidence to the contrary, that the images on our retinas uniquely determine out perceptual experiences. They are taking the camera analogy too far.

Having said all this, let me try to make clear what I do *not* mean to be claiming in this section, lest I be taken to be arguing for more than I intend to be. First, I am certainly not claiming that the physical causes of the images on our retinas have nothing to do with what we see. We cannot see just what we like. However, although the images on our retinas form part of the cause of what we see, another very important part of the cause is the inner state of our minds or brains, which will itself depend on our cultural upbringing, our knowledge and our expectations, and will not be determined solely by the physical properties of our eyes and the scene observed. Second, under a wide variety of circumstances, what we see in various situations remains fairly stable. The dependence of what we see on the state of our minds or brains is not so sensitive as to make communication, and science, impossible. Third, in all the examples quoted here, there is a sense in which all observers see the same thing. I accept and presuppose throughout this book that a single, unique, physical world exists independently of observers. Hence, when a number of observers look at a picture, a piece of apparatus, a microscope slide or whatever, there is a sense in which they are confronted by, look at, and hence see, the same thing. But it does not follow from this that they have identical perceptual experiences. There is a very important sense in which they do not see the same thing, and it is that latter sense on which I base some of my queries concerning the view that facts are unproblematically and directly given to observers through the senses. To what extent this undermines the view that facts adequate for science can be established by the senses remains to be seen.

Observable facts expressed as statements

In normal linguistic usage, the meaning of 'fact' is ambiguous. It can refer to a statement that expresses the fact and it can also refer to the state of affairs referred to by such a statement. For example, it is a fact that there are mountains and craters on the moon. Here the fact can be taken as referring to the mountains or craters themselves. Alternatively, the statement 'there are mountains and craters on the moon' can be taken as constituting the fact. When it is claimed that science is based on and derived from the facts, it is clearly the latter interpretation that is appropriate. Knowledge about the moon's surface is not based on and derived from mountains and craters but from factual statements about mountains and craters.

As well as distinguishing facts, understood as statements, from the states of affairs described by those statements, it is also clearly necessary to distinguish statements of facts from the perceptions that might occasion the acceptance of those statements as facts. For example, it is undoubtedly the case that when Darwin underwent his famous voyage on the *Beagle* he encountered many novel species of plant and animal, and so was subject to a range of novel perceptual experiences. However, he would have made no significant contribution to science had he left it at that. It was only when he had formulated statements describing the novelties and made them available to other scientists that he made a significant contribution to biology. To the extent that the voyage on the *Beagle* yielded novel facts to which an evolutionary theory could be related, it was statements that constituted those facts. For those who wish to claim that knowledge is derived from facts, they must have statements in mind, and neither perceptions nor objects like mountains and craters.

With this clarification behind us, let us return to the claims (a) to (c) about the nature of facts which concluded the first section of this chapter. Once we do so they immediately become highly problematic as they stand. Given that the facts that might constitute a suitable basis for science must be in the form of statements, the claim that facts are given in a straightforward way via the senses begins to look quite misconceived. For even if we set aside the difficulties highlighted in the previous section, and assume that perceptions are straightforwardly given in the act of seeing, it is clearly not the case that statements describing observable states of affairs (I will call them observation statements) are given to observers via the senses. It is absurd to think that *statements* of fact enter the brain by way of the senses.

Before an observer can formulate and assent to an observation statement, he or she must be in possession of the appropriate conceptual framework and a knowledge of how to appropriately apply it. That this is so becomes clear when we contemplate the way in which a child learns to describe (that is, make factual statements about) the world. Think of a parent teaching a child to recognise and describe apples. The parent shows the child an apple, points to it, and utters the word 'apple'. The child soon learns to repeat the word 'apple' in imitation. Having mastered this particular accomplishment, perhaps on a later day the child encounters its sibling's tennis ball, points and says 'apple'. At this point the parent intervenes to explain that the ball is not an apple, demonstrating, for example, that one cannot bite it like an apple. Further mistakes by the child, such as the identification of a choko as an apple, will require somewhat more elaborate explanations from the parent. By the time the child can successfully say there is an apple present when there is one, it has learnt quite a lot about apples. So it would seem that it is a mistake to presume that we must first observe the facts about apples before deriving knowledge about them from those facts, because the appropriate facts, formulated as statements, presuppose quite a lot of knowledge about apples.

Let us move from talk of children to some examples that are more relevant to our task of understanding science. Imagine a skilled botanist accompanied by someone like myself who is largely ignorant of botany taking part in a field trip into the Australian bush, with the objective of collecting observable facts about the native flora. It is undoubtedly the case that the botanist will be capable of collecting facts that are far more numerous and discerning than those I am able to observe and formulate, and the reason is clear. The botanist has a more elaborate conceptual scheme to exploit than myself, and that is because he or she knows more botany than I do. A knowledge of botany is a prerequisite for the formulation of the observation statements that might constitute its factual basis.

Thus, the recording of observable facts requires more than the reception of the stimuli, in the form of light rays, that impinge on the eye. It requires the knowledge of the appropriate conceptual scheme and how to apply it. In this sense, assumptions (a) and (b) cannot be accepted as they stand. Statements of fact are not determined in a straightforward way by sensual stimuli, and observation statements presuppose knowledge, so it cannot be the case that we first establish the facts and then derive our knowledge from them.

Why should facts precede theory?

I have taken as my starting point a rather extreme interpretation of the claim that science is derived from the facts. I have taken it to imply that the facts must be established prior to the derivation of scientific knowledge from them. First establish the facts and then build your theory to fit them. Both the fact that our perceptions depend to some extent on our prior knowledge and hence on our state of preparedness and our expectations (discussed earlier in the chapter) and the fact that observation statements presuppose the appropriate conceptual framework (discussed in the previous section) indicate that it is a demand that is impossible to live up to. Indeed, once it is subject to a close inspection it is a rather silly idea, so silly that I doubt if any serious philosopher of science would wish to defend it. How can we establish significant facts about the world through observation if we do not have some guidance as to what kind of knowledge we are seeking or what problems we are trying to solve? In order to make observations that might make a significant contribution to botany, I need to know much botany to start with. What is more, the very idea that the adequacy of our scientific knowledge should be tested against the observable facts would make no sense if, in proper science, the relevant facts must always precede the knowledge that might be supported by them. Our search for relevant facts needs to be guided by our current state of knowledge, which tells us, for example, that measuring the ozone concentration at various locations in the atmosphere yields relevant facts, whereas measuring the average hair length of the youths in Sydney does not. So let us drop the demand that the acquisition of facts should come before the formulation of the laws and theories that constitute scientific knowledge, and see what we can salvage of the idea that science is based on facts once we have done so.

According to our modified stand, we freely acknowledge that the formulation of observation statements presupposes significant knowledge, and that the search for relevant observable facts in science is guided by that knowledge. Neither acknowledgment necessarily undermines the claim that knowledge has a factual basis established by observation. Let us first take the point that the formulation of significant observation statements presupposes knowledge of the appropriate conceptual framework. Here we note that the availability of the conceptual resources for formulating observation statements is one thing. The truth or falsity of those statements is another. Looking at my solid state physics textbook, I can extract two observation statements, 'the crystal structure of diamond has inversion symmetry' and 'in a crystal of zinc sulphide there are four molecules per unit cell'. A degree of knowledge about crystal structures and how they are characterised is necessary for the formulation and understanding of these statements. But even if you do not have that knowledge, you will be able to recognise that there are other, similar, statements that can be formulated using the same terms, statements such as 'the crystal structure of diamond does not have inversion symmetry' and 'the crystal of zinc sulphide has six molecules per unit cell'. All of these statements are observation statements in the sense that once one has mastered the appropriate observational techniques their truth or falsity can be established by observation. When this is done, only the statements I extracted from my textbook are confirmed by observation, while the alternatives constructed from them are refuted. This illustrates the point that the fact that knowledge is necessary for the formulation of significant observation statements still leaves open the question of which of the statements so formulated are borne out by observation and which are not. Consequently, the idea that knowledge should be based on facts that are confirmed by observation is not undermined by the recognition that the formulation of the statements describing those facts are knowledge-dependent. There is only a problem if one sticks to the silly demand that the confirmation of facts relevant to some body of knowledge should precede the acquisition of any knowledge.

The idea that scientific knowledge should be based on facts established by observation need not be undermined, then, by the acknowledgment that the search for and formulation of those facts are knowledge-dependent. If the truth or falsity of observation statements can be established in a direct way by observation, then, irrespective of the way in which those statements came to be formulated, it would seem that the observation statements confirmed in this way provide us with a significant factual basis for scientific knowledge.

The fallibility of observation statements

We have made some headway in our search for a characterisation of the observational base of science, but we are not out of trouble yet. In the previous section our analysis presupposed that the truth or otherwise of observation statements can be securely established by observation in an unproblematic way. But is such a presupposition legitimate? We have already seen ways in which problems can arise from the fact that different observers do not necessarily have the same perceptions when viewing the same scene, and this can lead to disagreements about what the observable states of affairs are. The significance of this point for science is borne out by well-documented cases in the history of science, such as the dispute about whether or not the effects of so-called N-rays are observable, described by Nye (1980), and the disagreement between Sydney and Cambridge astronomers over what the observable facts were in the early years of radio astronomy, as described by Edge and Mulkay (1976). We have as yet said little to show how a secure observational basis for science can be established in the face of such difficulties. Further difficulties concerning the reliability of the observational basis of science arise from some of the ways in which judgments about the adequacy of observation statements draw on presupposed knowledge in a way that renders those judgments fallible. I will illustrate this with examples.

Aristotle included fire among the four elements of which all terrestrial objects are made. The assumption that fire is a distinctive substance, albeit a very light one, persisted for hundreds of years, and it took modern chemistry to thoroughly undermine it. Those who worked with this presupposition considered themselves to be observing fire directly when watching flames rise into the air, so that for them 'the fire ascended' is an observation statement that was frequently borne out by direct observation. We now reject such observation statements. The point is that if the knowledge that provides the categories we use to describe our observations is defective, the observation statements that presuppose those categories are similarly defective.

My second example concerns the realisation, established in the sixteenth and seventeenth centuries, that the earth moves, spinning on its axis and orbiting the sun. Prior to the circumstances that made this realisation possible, it can be said that the statement 'the earth is stationary' was a fact confirmed by observation. After all, one cannot see or feel it move, and if we jump in the air, the earth does not spin away beneath us. We, from a modern perspective, know that the observation statement in question is false in spite of these appearances. We understand inertia, and know that if we are moving in a horizontal direction at over one hundred metres per second because the earth is spinning, there is no reason why that should change when we jump in the air. It takes a force to change speed, and, in our example, there are no horizontal forces acting. So we retain the horizontal speed we share with the earth's surface and land where we took off. 'The earth is stationary' is not established by the observable evidence in the way it was once thought to be. But to fully appreciate why this is so, we need to understand inertia. That understanding was a seventeenth-century innovation. We have an example that illustrates a way in which the judgment of the truth or otherwise of an observation statement depends on the knowledge that forms the background against which the judgment is made. It would seem that the scientific revolution involved not just a progressive transformation of scientific theory, but also a transformation in what were considered to be the observable facts!

This last point is further illustrated by my third example. It concerns the sizes of the planets Venus and Mars as viewed from earth during the course of the year. It is a consequence of Copernicus's suggestion that the earth circulates the sun, in an orbit outside that of Venus and inside that of Mars, that the apparent size of both Venus and Mars should change appreciably during the course of the year. This is because when the earth is around the same side of the sun as one of those planets it is relatively close to it, whereas when it is on the opposite side of the sun to one of them it is relatively distant from it. When the matter is considered quantitatively, as it can be within Copernicus's own version of his theory, the effect is a sizeable one, with a predicted change in apparent diameter by a factor of about eight in the case of Mars and about six in the case of Venus. However, when the planets are observed carefully with the naked eye, no change in size can be detected for Venus, and Mars changes in size by no more than a factor of two. So the observation statement 'the apparent size of Venus does not change size during the course of the year' was straightforwardly confirmed, and was referred to in the Preface to Copernicus's *On the Revolutions of the Heavenly Spheres* as a fact confirmed 'by all the experience of the ages' (Duncan, 1976, p. 22). Osiander, who was the author of the Preface in question, was so impressed by the clash between the consequences of the Copernican theory and our 'observable fact' that he used it to argue that the Copernican theory should not be taken literally. We now know that the naked-eye observations of planetary sizes are deceptive, and that the eye is a very unreliable device for gauging the size of small light sources against a dark background. But it took Galileo to point this out and to show how the predicted change in size can be clearly discerned if Venus and Mars are viewed through a telescope. Here we have a clear example of the correction of a mistake about the observable facts made possible by improved knowledge and technology. In itself the example is unremarkable and non-mysterious. But it does show that any view to the effect that scientific knowledge is based on the facts acquired by observation must allow that the facts as well as the knowledge are fallible and subject to correction and that scientific knowledge and the facts on which it might be said to be based are interdependent.

The intuition that I intended to capture with my slogan 'science is derived from the facts' was that scientific knowledge has a special status in part because it is founded on a secure basis, solid facts firmly established by observation. Some of the considerations of this chapter pose a threat to this comfortable view. One difficulty concerns the extent to which perceptions are influenced by the background and expectations of the observer, so that what appears to be an observable fact for one need not be for another.

The second source of difficulty stems from the extent to which judgments about the truth of observation statements depend on what is already known or assumed, thus rendering the observable facts as fallible as the presuppositions underlying them. Both kinds of difficulty suggest that maybe the observable basis for science is not as straightforward and secure as is widely and traditionally supposed. In the next chapter I try to mitigate these fears to some extent by considering the nature of observation, especially as it is employed in science, in a more discerning way than has been involved in our discussion up until now.

Further reading

For a classic discussion of how knowledge is seen by an empiricist as derived from what is delivered to the mind via the senses, see Locke (1967), and by a logical positivist, see Ayer (1940). Hanfling (1981) is an introduction to logical positivism generally, including its account of the observational basis of science. A challenge to these views at the level of perception is Hanson (1958, chapter 1). Useful discussions of the whole issue are to be found in Brown (1977) and Barnes, Bloor and Henry (1996, chapters 1–3).

CHAPTER 2

Observation as practical intervention

Observation: passive and private or active and public?

A common way in which observation is understood by a range of philosophers is to see it as a passive, private affair. It is passive insofar as it is presumed that when seeing, for example, we simply open and direct our eyes, let the information flow in, and record what is there to be seen. It is the perception itself in the mind or brain of the observer that is taken to directly validate the fact, which may be 'there is a red tomato in front of me' for example. If it is understood in this way, then the establishment of observable facts is a very private affair. It is accomplished by the individual closely attending to what is presented to him or her in the act of perception. Since two observers do not have access to each other's perceptions, there is no way they can enter into a dialogue about the validity of the facts they are presumed to establish.

This view of perception or observation, as passive and private, is totally inadequate, and does not give an accurate account of perception in everyday life, let alone science. Everyday observation is far from passive. There is a range of things that are *done*, many of them automatically and perhaps unconsciously, to establish the validity of a perception. In the act of seeing we scan objects, move our heads to test for expected changes in the observed scene and so on. If we are not sure whether a scene viewed through a window is something out of the window or a reflection in the window, we can move our heads to check for the effect this has on the direction in which the scene is visible. It is a general point that if for any reason we doubt the validity of what seems to be the case on the basis of our perceptions, there are various actions we can take to remove the problem. If, in the example above, we have reason to suspect that the image of the tomato is some cleverly contrived optical image rather than a real tomato, we can touch it as well as look at it, and if necessary we can taste it or dissect it.

With these few, somewhat elementary, observations I have only touched the surface of the detailed story psychologists can tell about the range of things that are done by individuals in the act of perception. More important for our task is to consider the significance of the point for the role of observation in science. An example that illustrates my point well is drawn from early uses of the microscope in science. When scientists such as Robert Hooke and Henry Power used the microscope to look at small insects such as flies and ants, they often disagreed about the observable facts, at least initially. Hooke traced the cause of some of the disagreements to different kinds of illumination. He pointed out that the eye of a fly appears like a lattice covered with holes in one kind of light (which, incidentally, seems to have led Power to believe that this was indeed the case), like a surface covered with cones in another and in yet another light like a surface covered with pyramids. Hooke proceeded to make practical interventions designed to clear up the problem. He endeavoured to eliminate spurious information arising from dazzle and complicated reflections by illuminating specimens uniformly. He did this by using for illumination the light of a candle diffused through a solution of brine. He also illuminated his specimens from various directions to determine which features remained invariant under such changes. Some of the insects needed to be thoroughly intoxicated with brandy to render them both motionless and undamaged.

Hooke's book, *Micrographia* (1665), contains many detailed descriptions and drawings that resulted from Hooke's actions and observations. These productions were and are public, not private. They can be checked, criticised and added to by others. If a fly's eye, in some kinds of illumination, appears to be covered with holes, then that state of affairs cannot be usefully evaluated by the observer closely attending to his or her perceptions. Hooke showed what could be *done* to check the authenticity of the appearances in such cases, and the procedures he recommended could be carried out by anyone suitably inclined or skilled. The observable facts about the structure of a fly's eye that eventuate result from a process that is both active and public.

The point that action can be taken to explore the adequacy of claims put forward as observable facts has the consequence that subjective aspects of perception need not be an intractable problem for science. Ways in which perceptions of the same scene can vary from observer to observer depending on their background, culture and expectations were discussed in the previous chapter. Problems that eventuate from this undoubted fact can be countered to a large extent by taking appropriate action. It should be no news to anyone that the perceptual judgments of individuals can be unreliable for a range of reasons. The challenge, in science, is to arrange the observable situation in such a way that the reliance on such judgments is minimised if not eliminated. An example will illustrate the point.

The moon illusion is a common phenomenon. When it is high in the sky, the moon appears much smaller than when it is low on the horizon. This is an illusion. The moon does not change size nor does its distance from earth alter during the few hours that it takes for its relative position to undergo the required change. However, we do not have to put our trust in subjective judgments about the moon's size. We can, for example, mount a sighting tube fitted with cross-wires in such a way that its orientation can be read on a scale. The angle subtended by the moon at the place of sighting can be determined by aligning the cross-wires with each side of the moon in turn and noting the difference in the corresponding scale readings. This can be done when the moon is high in the sky and repeated when it is near the horizon. The fact that the apparent size of the moon has remained unchanged is reflected in the fact that there is no significant variation in the differences between the scale readings in the two cases.

Galileo and the moons of Jupiter

In this section the relevance of the discussion in the previous section is illustrated with a historical example. Late in 1609 Galileo constructed a powerful telescope and used it to look at the heavens. Many of the novel observations he made in the ensuing three months were controversial, and very relevant to the astronomical debate concerning the validity of the Copernican theory, of which Galileo became an avid champion. Galileo claimed, for instance, to have sighted four moons orbiting the planet Jupiter, but he had trouble convincing others of the validity of his observations. The matter was of some moment. The Copernican theory involved the controversial claim that the earth moves, spinning on its axis once a day and orbiting the sun once a year. The received view that Copernicus had challenged in the first half of the previous century was that the earth is stationary, with the sun and planets orbiting it. One of the many, far from trivial, arguments against the motion of the earth was that, if it orbited the sun as Copernicus claimed, the moon would be left behind. This argument is undermined once it is acknowledged that Jupiter has moons. For even the opponents of Copernicus agreed that Jupiter moves. Consequently, any moons it has are carried with it, exhibiting the very phenomenon that the opponents of Copernicus claimed to be impossible in the case of the earth.

Whether Galileo's telescopic observations of moons around Jupiter were valid was a question of some moment then. In spite of the initial skepticism, and the apparent inability of a range of his contemporaries to discern the moons through the telescope, Galileo had convinced his rivals within a period of two years. Let us see how he was able to achieve that – how he was able to 'objectify' his observations of Jupiter's moons.

Galileo attached a scale, marked with equally spaced horizontal and vertical lines, to his telescope by a ring in such a way that the scale was face-on to the observer and could be slid up and down the length of the telescope. A viewer looking through the telescope with one eye could view the scale with the other. Sighting of the scale was facilitated by illuminating it with a small lamp. With the telescope trained on Jupiter, the scale was slid along the telescope until the image of Jupiter viewed through the telescope with one eye lay in the central square of the scale viewed with the other eye. With this accomplished, the position of a moon viewed through the telescope could be read on the scale, the reading corresponding to its distance from Jupiter in multiples of the diameter of Jupiter. The diameter of Jupiter was a convenient unit, since employing it as a standard automatically allowed for the fact that its apparent diameter as viewed from earth varies as that planet approaches and recedes from the earth.

Using this technique, Galileo was able to record the daily histories of the four 'starlets' accompanying Jupiter. He was able to show that the data were consistent with the assumption that the starlets were indeed moons orbiting Jupiter with a constant period. The assumption was borne out, not only by the quantitative measurements but also by the more qualitative observation that the satellites occasionally disappeared from view as they passed behind or in front of the parent planet or moved into its shadow.

Galileo was in a strong position to argue for the veracity of his observations of Jupiter's moons, in spite of the fact that they were invisible to the naked eye. He could, and did, argue against the suggestion that they were an illusion produced by the telescope by pointing out that that suggestion made it difficult to explain why the moons appeared near Jupiter and nowhere else. Galileo could also appeal to the consistency and repeatability of his measurements and their compatibility with the assumption that the moons orbit Jupiter with a constant period. Galileo's quantitative data were verified by independent observers, including observers at the Collegio Romano and the Court of the Pope in Rome who were opponents of the Copernican theory. What is more, Galileo was able to predict further positions of the moons and the occurrence of transits and eclipses, and these too were confirmed by himself and independent observers, as documented by Stillman Drake (1978, pp. 175–6, 236–7).

The veracity of the telescopic sightings was soon accepted by those of Galileo's contemporaries who were competent observers, even by those who had initially opposed him. It is true that some observers could never manage to discern the moons, but I suggest that this is of no more significance than the inability of James Thurber (1933, pp. 101–3) to discern the structure of plant cells through a microscope. The strength of Galileo's case for the veracity of his telescopic observations of the moons of Jupiter derives from the range of practical, objective tests that his claims could survive. Although his case might have stopped short of being absolutely conclusive, it was incomparably stronger than any that could be made for the alternative, namely, that his sightings were illusions or artifacts brought about by the telescope.

Observable facts objective but fallible

An attempt to rescue a reasonably strong version of what constitutes an observable fact from the criticisms that we have levelled at that notion might go along the following lines. An observation statement constitutes a fact worthy of forming part of the basis for science if it is such that it can be straightforwardly tested by the senses and withstands those tests. Here the 'straightforward' is intended to capture the idea that candidate observation statements should be such that their validity can be tested in ways that involve routine, objective procedures that do not necessitate fine, subjective judgments on the part of the observer. The emphasis on tests brings out the active, public character of the vindication of observation statements. In this way, perhaps we can capture a notion of fact unproblematically established by observation. After all, only a suitably addicted philosopher will wish to spend time doubting that such things as meter readings can be securely established, within some small margin of error, by careful use of the sense of sight.

A small price has to be paid for the notion of an observable fact put forward in the previous paragraph. That price is that observable facts are to some degree fallible and subject to revision. If a statement qualifies as an observable fact because it has passsed all the tests that can be levelled at it hitherto, this does not mean that it will necessarily survive new kinds of tests that become possible in the light of advances in knowledge and technology. We have already met two significant examples of observation statements that were accepted as facts on good grounds but were eventually rejected in the light of such advances, namely, 'the earth is stationary' and 'the apparent size of Mars and Venus does not change appreciably during the course of the year'.

According to the view put forward here, observations suitable for constituting a basis for scientific knowledge are both objective and fallible. They are objective insofar as they can be publicly tested by straightforward procedures, and they are fallible insofar as they may be undermined by new kinds of tests made possible by advances in science and technology. This point can be illustrated by another example from the work of Galileo. In his *Dialogue Concerning the Two Chief World Systems* (1967, pp. 361–3) Galileo described an objective method for measuring the diameter of a star. He hung a cord between himself and the star at a distance such that the cord just blocked out the star. Galileo argued that the angle subtended at the eye by the cord was then equal to the angle subtended at the eye by the star. We now know that Galileo's results were spurious. The apparent size of a star as perceived by us is due entirely to atmospheric and other noise effects and has no determinate relation to the star's physical size. Galileo's measurements of star-size rested on implicit assumptions that are now rejected. But this rejection has nothing to do with subjective aspects of perception. Galileo's observations were objective in the sense that they involved routine procedures which, if repeated today, would give much the same results as obtained by Galileo. In the next chapter we will have cause to develop further the point that the lack of an infallible observational base for science does not derive solely from subjective aspects of perception.

Further reading

For a classic discussion of the empirical basis of science as those statements that withstand tests, see Popper (1972, chapter 5). The active aspects of observation are stressed in the second half of Hacking (1983), in Popper (1979, pp. 341–61) and in Chalmers (1990, chapter 4). Also of relevance is Shapere (1982).

CHAPTER 3

Experiment

Not just facts but relevant facts

In this chapter I assume for the sake of argument that secure facts can be established by careful use of the senses. After all, as I have already suggested, there is a range of situations relevant to science where this assumption is surely justified. Counting clicks on a Geiger counter and noting the position of a needle on a scale are unproblematic examples. Does the availability of such facts solve our problem about the factual basis for science? Do the statements that we assume can be established by observation constitute the facts from which scientific knowledge can be derived? In this chapter we will see that the answer to these questions is a decisive 'no'.

One point that should be noted is that what is needed in science is not just facts but relevant facts. The vast majority of facts that can be established by observation, such as the number of books in my office or the colour of my neighbour's car, are totally irrelevant for science, and scientists would be wasting their time collecting them. Which facts are relevant and which are not relevant to a science will be relative to the current state of development of that science. Science poses the questions, and ideally observation can provide answers. This is part of the answer to the question of what constitutes a relevant fact for science.

However, there is a more substantial point to be made, which I will introduce with a story. When I was young, my brother and I disagreed about how to explain the fact that the grass grows longer among the cow pats in a field than elsewhere in the same field, a fact that I am sure we were not the first to notice. My brother was of the opinion that it was the fertilising effect of the dung that was responsible, whereas I suspected that it was a mulching effect, the dung trapping moisture beneath it and inhibiting evaporation. I now have a strong suspicion that neither of us was entirely right and that the main explanation is simply that cows are disinclined to eat the grass around their own dung. Presumably all three of these effects play some role, but it is not possible to sort out the relative magnitudes of the effects by observations of the kind made by my brother and me. Some intervention would be necessary, such as, for example, locking the cows out of a field for a season to see if this reduced or eliminated the longer growth among the cow pats, by grinding the dung in such a way that the mulching effect is eliminated but the fertilising effect retained, and so on.

The situation exemplified here is typical. Many kinds of processes are at work in the world around us, and they are all superimposed on, and interact with, each other in complicated ways. A falling leaf is subject to gravity, air resistance and the force of winds and will also rot to some small degree as it falls. It is not possible to arrive at an understanding of these various processes by careful observation of events as they typically and naturally occur. Observation of falling leaves will not yield Galileo's law of fall. The lesson to be learnt here is rather straightforward. To acquire facts relevant for the identification and specification of the various processes at work in nature it is, in general, necessary to practically intervene to try to isolate the process under investigation and eliminate the effects of others. In short, it is necessary to do experiments.

It has taken us a while to get to this point, but it should perhaps be somewhat obvious that if there are facts that constitute the basis for science, then those facts come in the form of experimental results rather than any old observable facts. As obvious as this might be, it is not until the last couple of decades that philosophers of science have taken a close look at the nature of experiment and the role it plays in science. Indeed, it is an issue that was given little attention in the previous editions of this book. Once we focus on experiment rather than mere observation as supplying the basis for science, the issues we have been discussing take on a somewhat different light, as we shall see in the remainder of this chapter.

The production and updating of experimental results

Experimental results are by no means straightforwardly given. As any experimentalist, and indeed any science student, knows, getting an experiment to work is no easy matter. A significant new experiment can take months or even years to successfully execute. A brief account of my own experiences as an experimental physicist in the 1960s will illustrate the point nicely. It is of no great importance whether the reader follows the detail of the story. I simply aim to give some idea of the complexity and practical struggle involved in the production of an experimental result.

The aim of my experiment was to scatter low-energy electrons from molecules to find out how much energy they lost in the process, thereby gaining information related to the energy levels in the molecules themselves. To reach this objective, it was necessary to produce a beam of electrons that all moved at the same velocity and hence had the same energy. It was necessary to arrange for them to collide with one target molecule only before entering the detector, otherwise the sought-for information would be lost, and it was necessary to measure the velocity, or energy, of the scattered electrons with a suitably designed detector. Each of these steps posed a practical challenge. The velocity selector involved two conducting plates bent into concentric circles with a potential difference between them. Electrons entering between the plates would only emerge from the other end of the circular channel if they had a velocity that matched the potential difference between the plates. Otherwise they would be deflected onto the conducting plates. To ensure that the electrons were likely to collide with only one target molecule it was necessary to do the experiment in a region that was highly evacuated, containing a sample of the target gas at very low pressure. This required pushing the available vacuum technology to its limits. The velocity of scattered electrons was to be measured by an arrangement of circular electrodes similar to that used in producing the mono-energetic beam. The intensity of electrons scattered with a particular velocity could be measured by setting the potential difference between the plates to a value that allowed only the electrons with that velocity to traverse the circle and emerge at the other end of the analyser. Detecting the emerging electrons involved measuring a minutely small current that again pushed the available technology to its limits.

That was the general idea, but each step presented a range of practical problems of a sort that will be familiar to anyone who has worked in this kind of field. It was very difficult to rid the apparatus of unwanted gases that were emitted from the various metals from which the apparatus was made. Molecules of these gases that were ionised by the electron beam could coagulate on the electrodes and cause spurious electric potentials. Our American rivals found that gold-plating the electrodes helped greatly to minimise these problems. We found that coating them with a carbon-based solvent called 'aquadag' was a big help, not quite as effective as gold-plating but more in keeping with our research budget. My patience (and my research scholarship) ran out well before this experiment was made to yield significant results. I understand that a few more research students came to grief before significant results were eventually obtained. Now, decades later, low-energy electron spectroscopy is a pretty standard technique.

The details of my efforts, and those of my successors who were more successful, are not important. What I have said should be sufficient to illustrate what should be an uncontentious point. If experimental results constitute the facts on which science is based, then they are certainly not straightforwardly given via the senses. They have to be worked for, and their establishment involves considerable know-how and practical trial and error as well as exploitation of the available technology.

Nor are judgments about the adequacy of experimental results straightforward. Experiments are adequate, and interpretable as displaying or measuring what they are intended to display or measure, only if the experimental set-up is appropriate and disturbing factors have been eliminated. This in turn will require that it is known what those disturbing factors are and how they can be eliminated. Any inadequacies in the relevant knowledge about these factors could lead to inappropriate experimental measures and faulty conclusions. So there is a significant sense in which experimental facts and theory are interrelated. Experimental results can be faulty if the knowledge informing them is deficient or faulty.

A consequence of these general, and in a sense quite mundane, features of experiment is that experimental results are fallible, and can be updated or replaced for reasonably straightforward reasons. Experimental results can become outmoded because of advances in technology, they can be rejected because of some advance in understanding (in the light of which an experimental set-up comes to be seen as inadequate) and they can be ignored as irrelevant in the light of some shift in theoretical understanding. These points and their significance are illustrated by historical examples in the next section.

Transforming the experimental base of science: historical examples

Discharge tube phenomena commanded great scientific interest in the final quarter of the nineteenth century. If a high voltage is connected across metal plates inserted at each end of an enclosed glass tube, an electric discharge occurs, causing various kinds of glowing within the tube. If the gas pressure within the tube is not too great, streamers are produced, joining the negative plate (the cathode) and the positive plate (the anode). These became known as cathode rays, and their nature was a matter of considerable interest to scientists of the time. The German physicist Heinrich Hertz conducted a series of experiments in the early 1880s intended to shed light on their nature. As a result of these experiments Hertz concluded that cathode rays are not beams of charged particles. He reached this conclusion in part because the rays did not seem to be deflected when they were subjected to an electric field perpendicular to their direction of motion as would be expected of a beam of charged particles. We now regard Hertz's conclusion as false and his experiments inadequate. Before the century had ended, J. J. Thomson had conducted experiments that showed convincingly that cathode rays are deflected by electric and magnetic fields in a way that is consistent with their being beams of charged particles and was able to measure the ratio of the electric charge to the mass of the particles.

It was improved technology and improved understanding of the situation that made it possible for Thomson to improve on and reject Hertz's experimental results. The electrons that constitute the cathode rays can ionise the molecules of the gas in the tube, that is, displace an electron or two from them so that they become positively charged. These ions can collect on metal plates in the apparatus and lead to what, from the point of view of the experiments under consideration, are spurious electric fields. It was presumably such fields that prevented Hertz producing the deflections that Thomson was eventually to be able to produce and measure. The main way that Thomson was able to improve on Hertz's efforts was to take advantage of improved vacuum technology to remove more gas molecules from the tube. He subjected his apparatus to prolonged baking to drive residual gas from the various surfaces within the tube. He ran the vacuum pump for several days to remove as much of the residual gas as possible. With an improved vacuum, and with a more appropriate arrangement of electrodes, Thomson was able to establish the deflections that Hertz had declared to be non-existent. When Thomson allowed the pressure in his apparatus to rise to what it had been in Hertz's, Thomson could not detect a deflection either. It is important to realise here that Hertz is not to be blamed for drawing the conclusion that he did. Given his understanding of the situation, and drawing on the knowledge available to him, he had good reasons to believe that the pressure in his apparatus was sufficiently low and that his apparatus was appropriately arranged. It was only in the light of subsequent theoretical and technological advances that his experiment came to be seen as deficient. The moral, of course, is this: who knows which contemporary experimental results will be shown to be deficient by advances that lie ahead?

Far from being a shoddy experimentalist, the fact that Hertz was one of the very best is borne out by his success in being the first to produce radio waves in 1888, as the culmination of two years of brilliant experimental research. Apart from revealing a new phenomenon to be explored and developed experimentally, Hertz's waves had considerable theoretical significance, since they confirmed Maxwell's electromagnetic theory, which he had formulated in the mid-1860s and which had the consequence that there be such waves (although Maxwell himself had not realised this). Most aspects of Hertz's results remain acceptable and retain their significance today. However, some of his results needed to be replaced and one of his main interpretations of them rejected. Both of these points illustrate the way in which experimental results are subject to revision and improvement.

Hertz was able to use his apparatus to generate standing waves, which enabled him to measure their wavelength, from which he could deduce their velocity. His results indicated that the waves of longer wavelength travelled at a greater speed in air than along wires, and faster than light, whereas Maxwell's theory predicted that they would travel at the speed of light both in air and along the wires of Hertz's apparatus. The results were inadequate for reasons that Hertz already suspected. Waves reflected back onto the apparatus from the walls of the laboratory were causing unwanted interference. Hertz (1962, p. 14) himself reflected on the results as follows:

> The reader may perhaps ask why I have not endeavored to settle the doubtful point myself by repeating the experiments. I have indeed repeated the experiments, but have only found, as might be expected, that a simple repetition under the same conditions cannot remove the doubt, but rather increases it. A definite decision can only be arrived at by experiments carried out under more favorable conditions. More favorable conditions here mean larger rooms, and such were not at my disposal. I again emphasise the statement that care in making the observations cannot make up for want of space. If the long waves cannot develop, they clearly cannot be observed.

Hertz's experimental results were inadequate because his experimental set-up was inappropriate for the task in hand. The wavelengths of the waves investigated needed to be small compared with the dimensions of the laboratory if unwanted interference from reflected waves was to be removed. As it transpired, within a few years experiments were carried out 'under more favorable conditions' and yielded velocities in line with the theoretical predictions.

A point to be stressed here is that experimental results are required not only to be adequate, in the sense of being accurate recordings of what happened, but also to be appropriate or significant. They will typically be designed to cast light on some significant question. Judgments about what is a significant question and about whether some specific set of experiments is an adequate way of answering it will depend heavily on how the practical and theoretical situation is understood. It was the existence of competing theories of electromagnetism and the fact that one of the major contenders predicted radio waves travelling with the speed of light that made Hertz's attempt to measure the velocity of his waves particularly significant, while it was an understanding of the reflection behaviour of the waves that led to the appreciation that Hertz's experimental set-up was inappropriate. These particular results of Hertz's were rejected and soon replaced for reasons that are straightforward and non-mysterious from the point of view of physics.

As well as illustrating the point that experiments need to be appropriate or significant, and that experimental results are replaced or rejected when they cease to be so, this episode in Hertz's researches and his own reflections on it clearly bring out the respect in which the rejection of his velocity measurements has nothing whatsoever to do with problems of human perception. There is no reason whatsoever to doubt that Hertz carefully observed his apparatus, measuring distances, noting the presence or absence of sparks across the gaps in his detectors, and recording instrument readings. His results can be assumed to be objective in the sense that anyone who repeats them will get similar results. Hertz himself stressed this point. The problem with Hertz's experimental results stems neither from inadequacies in his observations nor from any lack of repeatability, but rather from the inadequacy of the experimental set-up. As Hertz pointed out, 'care in making the observations cannot make up for want of space'. Even if we concede that Hertz was able to establish secure facts by way of careful observation, we can see that this in itself was insufficient to yield experimental results adequate for the scientific task in question.

The above discussion can be construed as illustrating how the acceptability of experimental results is theory-dependent, and how judgments in this respect are subject to change as our scientific understanding develops. This is illustrated at a more general level by the way in which the significance of Hertz's production of radio waves has changed since Hertz first produced them. At that time, one of the several competing theories of electromagnetism was that of James Clerk Maxwell, who had developed the key ideas of

Michael Faraday and had understood electric and magnetic states as the mechanical states of an all-pervasive ether. This theory, unlike its competitors, which assumed that electric currents, charges and magnets acted on each other at a distance and did not involve an ether, predicted the possibility of radio waves moving at the speed of light. This is the aspect of the state of development of physics that gave Hertz's results their theoretical significance. Consequently, Hertz and his contemporaries were able to construe the production of radio waves as, among other things, *confirmation of the existence of an ether.* Two decades later the ether was dispensed with in the light of Einstein's special theory of relativity. Hertz's results are still regarded as confirming Maxwell's theory, but only a rewritten version of it that dispenses with the ether, and treats electric and magnetic fields as real entities in their own right.

Another example, concerning nineteenth-century measurements of molecular weights, further illustrates the way in which the relevance and interpretation of experimental results depend on the theoretical context. Measurements of the molecular weights of naturally occurring elements and compounds were considered to be of fundamental importance by chemists in the second half of the nineteenth century in the light of the atomic theory of chemical combination. This was especially so for those who favoured Prout's hypothesis that the hydrogen atom is the basic building block from which other atoms are constructed, for this led one to expect that molecular weights measured relative to hydrogen would be whole numbers. The painstaking measurements of molecular weights by the leading experimental chemists in the nineteenth century became largely irrelevant from the point of view of theoretical chemistry once it was realised that naturally occurring elements contain a mixture of isotopes in proportions that had no particular theoretical significance. This situation inspired the chemist F. Soddy to comment on its outcome as follows (Lakatos and Musgrave, 1970, p. 140):

> There is something surely akin to if not transcending tragedy in the fate that has overtaken the life work of this distinguished galaxy of nineteenth-century chemists, rightly revered by their contemporaries as representing the crown and perfection of accurate scientific measurements. Their hard won results, for the

> moment at least, appear as of little significance as the determination of the average weight of a collection of bottles, some of them full and some of them more or less empty.

Here we witness old experimental results being set aside as irrelevant, and for reasons that do not stem from problematic features of human perception. The nineteenth-century chemists involved were 'revered by their contemporaries as representing the crown and perfection of accurate scientific measurement' and we have no reason to doubt their observations. Nor need we doubt the objectivity of the latter. I have no doubt that similar results would be obtained by contemporary chemists if they were to repeat the same experiments. That they be adequately performed is a necessary but not sufficient condition for the acceptability of experimental results. They need also to be relevant and significant.

The points I have been making with the aid of examples can be summed up in a way that I believe is quite uncontentious from the point of view of physics and chemistry and their practice. The stock of experimental results regarded as an appropriate basis for science is constantly updated. Old experimental results are rejected as inadequate and replaced by more adequate ones, for a range of fairly straightforward reasons. They can be rejected because the experiment involved inadequate precautions against possible sources of interference, because the measurements employed insensitive and outmoded methods of detection, because the experiments came to be understood as incapable of solving the problem in hand, or because the questions they were designed to answer became discredited. Although these observations can be seen as fairly obvious comments on everyday scientific activity, they nevertheless have serious implications for much orthodox philosophy of science, for they undermine the widely held notion that science rests on secure foundations. What is more, the reasons why it does not has nothing much to do with problematic features of human perception.

Experiment as an adequate basis for science

In the previous sections of this chapter I have subjected to critical scrutiny the idea that experimental results are straightforwardly given and totally secure. I have made a case to the effect that they are theory-dependent in certain respects and fallible and revisable. This can be interpreted as a serious threat to the idea that scientific knowledge is special because it is supported by experience in some especially demanding and convincing way. If, it might be argued, the experimental basis of science is as fallible and revisable as I have argued it to be, then the knowledge based on it must be equally fallible and revisable. The worry can be strengthened by pointing to a threat of circularity in the way scientific theories are alleged to be borne out by experiment. If theories are appealed to in order to judge the adequacy of experimental results, and those same experimental results are taken as the evidence for the theories, then it would seem that we are caught in a circle. It would seem that there is a strong possibility that science will not provide the resources to settle a dispute between the proponents of opposing theories by appeal to experimental results. One group would appeal to its theory to vindicate certain experimental results, and the opposing camp would appeal to its rival theory to vindicate different experimental results. In this section I give reasons for resisting these extreme conclusions.

It must be acknowledged that there is the possibility that the relationship between theory and experiment might involve a circular argument. This can be illustrated by the following story from my schoolteaching days. My pupils were required to conduct an experiment along the following lines. The aim was to measure the deflection of a current-carrying coil suspended between the poles of a horseshoe magnet and free to rotate about an axis perpendicular to the line joining the poles of the magnet. The coil formed part of a circuit containing a battery to supply a current, an ammeter to measure the current and a variable resistance to make it possible to adjust the strength of the current. The aim was to note the deflection of the magnet corresponding to various values of the current in the circuit as registered by the ammeter. The experiment was to be deemed a success for those pupils who got a nice straight-line graph when they plotted deflection against current, revealing the proportionality of the two. I remember being disconcerted by this experiment, although, perhaps wisely, I did not transmit my worry to my pupils. My worry stemmed from the fact that I knew what was inside the ammeter. What was inside was a coil suspended between the poles of a magnet in such a way that it was deflected by a current through the coil causing a needle to move on the visible and evenly calibrated scale of the ammeter. In this experiment, then, the proportionality of deflection to current was already presupposed when the reading of the ammeter was taken as a measure of the current. What was taken to be supported by the experiment was already presupposed in it, and there was indeed a circularity.

My example illustrates how circularity can arise in arguments that appeal to experiment. But the very same example serves to show that this need not be the case. The above experiment could have, and indeed should have, used a method of measuring the current in the circuit that did not employ the deflection of a coil in a magnetic field. All experiments will presume the truth of some theories to help judge that the set-up is adequate and the instruments are reading what they are meant to read. But these presupposed theories need not be identical to the theory under test, and it would seem reasonable to assume that a prerequisite of good experimental design is to ensure that they are not.

Another point that serves to get the 'theory-dependence of experiment' in perspective is that, however informed by theory an experiment is, there is a strong sense in which the results of an experiment are determined by the world and not by the theories. Once the apparatus is set up, the circuits completed, the switches thrown and so on, there will or will not be a flash on the screen, the beam may or may not be deflected, the reading on the ammeter may or may not increase. We cannot make the outcomes conform to our theories. It was because the physical world is the way it is that the experiment conducted by Hertz yielded no deflection of cathode rays and the modified experiment conducted by Thomson did. It was the material differences in the experimental arrangements of the two physicists that led to the differing outcomes, not the differences in the theories held by them. It is the sense in which experimental outcomes are determined by the workings of the world rather than by theoretical views about the world that provides the possibility of testing theories against the world. This is not to say that significant results are easily achievable and infallible, nor that their significance is always straightforward. But it does help to establish the point that the attempt to test the adequacy of scientific theories against experimental results is a meaningful quest. What is more, the history of science gives us examples of cases where the challenge was successfully met.

Further reading

The second half of Hacking (1983) was an important early move in the new interest philosophers of science have taken in experiment. Other explorations of the topic are Franklin (1986), Franklin (1990), Galison (1987) and Mayo (1996), although these detailed treatments will take on their full significance only in the light of chapter 13, on the 'new experimentalism'.

CHAPTER 4

Deriving theories from the facts: induction

Introduction

In these early chapters of the book we have been considering the idea that what is characteristic of scientific knowledge is that it is derived from the facts. We have reached a stage where we have given some detailed attention to the nature of the observational and experimental facts that can be considered as the basis from which scientific knowledge might be derived, although we have seen that those facts cannot be established as straightforwardly and securely as is commonly supposed. Let us assume, then, that appropriate facts can be established in science. We must now face the question of how scientific knowledge can be derived from those facts.

'Science is derived from the facts' could be interpreted to mean that scientific knowledge is constructed by first establishing the facts and then subsequently building the theory to fit them. We discussed this view in chapter 1 and rejected it as unreasonable. The issue that I wish to explore involves interpreting 'derive' in some kind of logical rather than temporal sense. No matter which comes first, the facts or the theory, the question to be addressed is the extent to which the theory is borne out by the facts. The strongest possible claim would be that the theory can be logically derived from the facts. That is, given the facts, the theory can be proven as a consequence of them. This strong claim cannot be substantiated. To see why this is so we must look at some of the basic features of logical reasoning.

Baby logic

Logic is concerned with the deduction of statements from other, given, statements. It is concerned with what follows from what. No attempt will be made to give a detailed account and appraisal of logic or deductive reasoning here. Rather, I will make the points that will be sufficient for our purpose with the aid of some very simple examples.

Here is an example of a logical argument that is perfectly adequate or, to use the technical term used by logicians, perfectly valid.

Example 1
1. All books on philosophy are boring.
2. This book is a book on philosophy.

---

3. This book is boring.

In this argument, (1) and (2) are the premises and (3) is the conclusion. It is evident, I take it, that if (1) and (2) are true then (3) is bound to be true. It is not possible for (3) to be false once it is given that (1) and (2) are true. To assert (1) and (2) as true and to deny (3) is to contradict oneself. This is the key feature of a *logically valid* deduction. If the premises are true then the conclusion must be true. Logic is truth preserving.

A slight modification of Example 1 will give us an instance of an argument that is not valid.

Example 2
1. Many books on philosophy are boring.
2. This book is a book on philosophy.

---

3. This book is boring.

In this example, (3) does not follow of necessity from (1) and (2). Even if (1) and (2) are true, then this book might yet turn out to be one of the minority of books on philosophy that are not boring. Accepting (1) and (2) as true and holding (3) to be false does not involve a contradiction. The argument is invalid.

The reader may by now be feeling bored. Experiences of that kind certainly have a bearing on the truth of statements (1) and (3) in Example 1 and Example 2. But a point that needs to be stressed here is that logical deduction alone cannot establish the truth of factual statements of the kind figuring in our examples. All that logic can offer in this connection is that *if* the premises are true and the argument is valid *then* the conclusion must be true. But whether the premises are true or not is not a question that can be settled by an appeal to logic. An argument can be a perfectly valid deduction even if it involves a false premise. Here is an example.

Example 3
1. All cats have five legs.
2. Bugs Pussy is my cat.
____
3. Bugs Pussy has five legs.

This is a perfectly valid deduction. If (1) and (2) are true then (3) must be true. It so happens that, in this example (1) and (3) are false. But this does not affect the fact that the argument is valid.

There is a strong sense, then, in which logic alone is not a source of new truths. The truth of the factual statements that constitute the premises of arguments cannot be established by appeal to logic. Logic can simply reveal what follows from, or what in a sense is already contained in, the statements we already have to hand. Against this limitation we have the great strength of logic, namely, its truth-preserving character. If we can be sure our premises are true then we can be equally sure that everything we logically derive from them will also be true.

Can scientific laws be derived from the facts?

With this discussion of the nature of logic behind us, it can be straightforwardly shown that scientific knowledge cannot be derived from the facts if 'derive' is interpreted as 'logically deduce'.

Some simple examples of scientific knowledge will be sufficient for the illustration of this basic point. Let us consider some low-level scientific laws such as 'metals expand when heated' or 'acids turn litmus red'. These are general statements. They are examples of what philosophers refer to as universal statements. They refer to all events of a particular kind, all instances of metals being heated and all instances of litmus being immersed in acid. Scientific knowledge invariably involves general statements of this kind. The situation is quite otherwise when it comes to the observation statements that constitute the facts that provide the evidence for general scientific laws. Those observable facts or experimental results are specific claims about a state of affairs that obtains at a particular time. They are what philosophers call singular statements. They include statements such as 'the length of the copper bar increased when it was heated' or 'the litmus paper turned red when immersed in the beaker of hydrochloric acid'. Suppose we have a large number of such facts at our disposal as the basis from which we hope to derive some scientific knowledge (about metals or acids in the case of our examples). What kind of argument can take us from those facts, as premises, to the scientific laws we seek to derive as conclusions? In the case of our example concerning the expansion of metals the argument can be schematised as follows:

Premises
1. Metal x_1 expanded when heated on occasion t_1.
2. Metal x_2 expanded when heated on occasion t_2.
n. Metal x_n expanded when heated on occasion t_n.
Conclusion
All metals expand when heated.

This is not a logically valid argument. It lacks the basic features of such an argument. It is simply not the case that if the statements constituting the premises are true then the conclusion must be true. However many observations of expanding metals we have to work with, that is, however great n might be in our example, there can be no *logical* guarantee that some sample of metal might on some occasion contract when heated. There is no contradiction involved in claiming both that all known examples of the heating of metals has resulted in expansion and that 'all metals expand when heated' is false.

This straightforward point is illustrated by a somewhat gruesome example attributed to Bertrand Russell. It concerns a turkey who noted on his first morning at the turkey farm that he was fed at 9 am. After this experience had been repeated daily for several weeks the turkey felt safe in drawing the conclusuion 'I am always fed at 9 am'. Alas, this conclusion was shown to be false in no uncertain manner when, on Christmas eve, instead of being fed, the turkey's throat was cut. The turkey's argument led it from a number of true observations to a false conclusion, clearly indicating the invalidity of the argument from a logical point of view.

Arguments of the kind I have illustrated with the example concerning the expansion of metals, which proceed from a finite number of specific facts to a general conclusion, are called *inductive* arguments, as distinct from logical, *deductive* arguments. A characteristic of inductive arguments that distinguishes them from deductive ones is that, by proceeding as they do from statements about *some* to statements about *all* events of a particular kind, they go beyond what is contained in the premises. General scientific laws invariably go beyond the finite amount of observable evidence that is available to support them, and that is why they can never be proven in the sense of being logically deduced from that evidence.

What constitutes a good inductive argument?

We have seen that if scientific knowledge is to be understood as being derived from the facts, then 'derive' must be understood in an inductive rather than a deductive sense. But what are the characteristics of a good inductive argument? The question is of fundamental importance because it is clear that not all generalisations from the observable facts are warranted. Some of them we will wish to regard as overhasty or based on insufficient evidence, as when, perhaps, we condemn the attribution of some characteristic to an entire ethnic group based on some unpleasant encounters with just one pair of neighbours. Under precisely what circumstances is it legitimate to assert that a scientific law has been 'derived' from some finite body of observational and experimental evidence?

A first attempt at an answer to this question involves the demand that, if an inductive inference from observable facts to laws is to be justified, then the following conditions must be satisfied:

1. The number of observations forming the basis of a generalisation must be large.
2. The observations must be repeated under a wide variety of conditions.

3. No accepted observation statement should conflict with the derived law.

Condition 1 is regarded as necessary because it is clearly not legitimate to conclude that all metals expand when heated on the basis of just one observation of an iron bar's expansion, say, any more than it is legitimate to conclude that all Australians are drunkards on the basis of one observation of an intoxicated Australian. A large number of independent observations would appear to be necessary before either generalisation can be justified. A good inductive argument does not jump to conclusions.

One way of increasing the number of observations in the examples mentioned would be to repeatedly heat a single bar of metal or to continually observe a particular Australian getting drunk night after night, and perhaps morning after morning. Clearly, a list of observation statements acquired in such a way would form a very unsatisfactory basis for the respective generalisations. That is why Condition 2 is necessary. 'All metals expand when heated' will be a legitimate generalisation only if the observations of expansion on which it is based range over a wide variety of conditions. Various kinds of metals should be heated, long bars, short bars, silver bars, copper bats, etc. should be heated at high and low pressures and high and low temperatures and so on. Only if on all such occasions expansion results is it legitimate to generalise by induction to the general law. Further, it is evident that if a particular sample of metal is observed not to expand when heated, then the generalisation to the law will not be justified. Condition 3 is essential.

The above can be summed up by the following statement of *the principle of induction*.

> If a large number of **A**s have been observed under a wide variety of conditions, and if all those **A**s without exception possess the property **B**, then all **A**s have the property **B**.

There are serious problems with this characterisation of induction. Let us consider Condition 1, the demand for large numbers of observations. One problem with it is the vagueness of 'large'. Are a hundred, a thousand or more observations required? If we do attempt to introduce precision by introducing a number here, then there would surely be a great deal of arbitrariness in the number chosen. The problems do not stop here. There are many instances in which the demand for a large number of instances seems inappropriate. To illustrate this, consider the strong public reaction against nuclear warfare that was provoked by the dropping of the first atomic bomb on Hiroshima towards the end of the Second World War. That reaction was based on the realisation of the extent to which atomic bombs cause widespread destruction and human suffering. And yet this widespread, and surely reasonable, belief was based on just one dramatic observation. In a similar vein, it would be a very stubborn investigator who insisted on putting his hand in the fire many times before concluding that fire burns. Let us consider a less fanciful example related to scientific practice. Suppose I reproduced an experiment reported in some recent scientific journal, and sent my results off for publication. Surely the editor of the journal would reject my paper, explaining that the experiment had already been done! Condition 1 is riddled with problems.

Condition 2 has serious problems too, stemming from difficulties surrounding the question of what counts as a significant variation in circumstances. What counts as a significant variation in the circumstances under which the expansion of a heated metal is to be investigated? Is it necessary to vary the type of metal, the pressure and the time of day? The answer is 'yes' in the first and possibly the second case but 'no' in the third. But what are the grounds for that answer? The question is important because unless it can be answered the list of variations can be extended indefinitely by endlessly adding further variations, such as the size of the laboratory and the colour of the experimenter's socks. Unless such 'superfluous' variations can be eliminated, the conditions under which an inductive inference can be accepted can never be satisfied. What are the grounds, then, for regarding a range of possible variations as superfluous? The commonsense answer is straightforward enough. We draw on our prior knowledge of the situation to distinguish between the factors that might and those that cannot influence the system we are investigating. It is our knowledge of metals and the kinds of ways that they can be acted on that leads us to the expectation that their physical behaviour will depend on the type of metal and the surrounding pressure but not on the time of day or the colour of the experimenter's socks. We draw on our current stock of knowledge to help judge what is a relevant circumstance that might need to be varied when investigating the generality of an effect under investigation.

This response to the problem is surely correct. However, it poses a problem for a sufficiently strong version of the claim that scientific knowledge should be derived from the facts by induction. The problem arises when we pose the question of how the knowledge appealed to when judging the relevance or otherwise of some circumstances to a phenomenon under investigation (such as the expansion of metals) is itself vindicated. If we demand that that knowledge itself is to be arrived at by induction, then our problem will recur, because those further inductive arguments will themselves require the specification of the relevant circumstances and so on. Each inductive argument invovles an appeal to prior knowledge, which needs an inductive argument to justify it, which involves an appeal to further prior knowledge and so on in a never-ending chain. The demand that all knowledge be justified by induction becomes a demand that cannot be met.

Even Condition 3 is problematic since little scientific knowledge would survive the demand that there be no known exceptions. This is a point that will be discussed in some detail in chapter 7.

Further problems with inductivism

Let us call the position according to which scientific knowledge is to be derived from the observable facts by some kind of inductive inference *inductivism* and those who subscribe to that view *inductivists*. We have already pointed to a serious problem inherent in that view, namely, the problem of stating precisely under what conditions a generalisation constitutes a good inductive inference. That is, it is not clear what induction amounts to. There are further problems with the inductivist position.

If we take contemporary scientific knowledge at anything like face value, then it has to be admitted that much of that knowledge refers to the unobservable. It refers to such things as protons and electrons, genes and DNA molecules and so on. How can such knowledge be accommodated into the inductivist position? Insofar as inductive reasoning involves some kind of generalisation from observable facts, it would appear that such reasoning is not capable of yielding knowledge of the unobservable. Any generalisation from facts about the observable world can yield nothing other than generations about the observable world. Consequently, scientific knowledge of the unobservable world can never be established by the kind of inductive reasoning we have discussed. This leaves the inductivist in the uncomfortable position of having to reject much contemporary science on the grounds that it involves going beyond what can be justified by inductive generalisation from the observable.

Another problem stems from the fact that many scientific laws take the form of exact, mathematically formulated laws. The law of gravitation, which states that the force between any two masses is proportional to the product of those masses divided by the square of the distance that separates them, is a straightforward example. Compared with the exactness of such laws we have the inexactness of any of the measurements that constitute the observable evidence for them. It is well appreciated that all observations are subject to some degree of error, as reflected in the practice of scientists when they write the result of a particular measurement as $x \pm dx$, where dx represents the estimated margin of error. If scientific laws are inductive generalisations from observable facts it is difficult to see how one can escape the inexactness of the measurements that constitute the premises of the inductive arguments. It is difficult to see how exact laws can ever be inductively justified on the basis of inexact evidence.

A third problem for the inductivist is an old philosophical chestnut called the problem of induction. The problem arises for anyone who subscribes to the view that scientific knowledge in all its aspects must be justified either by an appeal to (deductive) logic or by deriving it from experience. David Hume was an eighteenth-century philosopher who did subscribe to that view, and it was he who clearly articulated the problem I am about to highlight.

The problem arises when we raise the question of how induction itself is to be justified. How is the principle of induction to be vindicated? Those who take the view under discussion have only two options, to justify it by an appeal to logic or by an appeal to experience. We have already seen that the first option will not do. Inductive inferences are not logical (deductive) inferences. This leaves us with the second option, to attempt to justify induction by an appeal to experience. What would such a justification be like? Presumably, it would go something like this. Induction has been observed to work on a large number of occasions. For instance, the laws of optics, derived by induction from the results of laboratory experiments, have been used on numerous occasions in the design of optical instruments that have operated satisfactorily, and the laws of planetary motion, inductively derived from the observation of planetary positions, have been successfully used to predict eclipses and conjunctions. This list could be greatly extended with accounts of successful predictions and explanations that we presume to be made on the basis of inductively derived scientific laws and theories. Thus, so the argument goes, induction is justified by experience.

This justification of induction is unacceptable. This can be seen once the form of the argument is spelt out schematically as follows:

> The principle of induction worked successfully on occasion x_1
> The principle of induction worked successfully on occasion x_2 etc.
> ---
> The principle of induction always works.

A general statement asserting the validity of the principle of induction is here inferred from a number of individual instances of its successful application. The argument is therefore itself an inductive one. Consequently, the attempt to justify induction by an appeal to experience involves assuming what one is trying to prove. It involves justifying induction by appealing to induction, and so is totally unsatisfactory.

One attempt to avoid the problem of induction involves weakening the demand that scientific knowledge be proven true, and resting content with the claim that scientific claims can be shown to be probably true in the light of the evidence. So the vast number of observations that can be invoked to support the claim that materials denser than air fall downwards on earth, although it does not permit us to prove the truth of the claim, does warrant the assertion that the claim is probably true. In line with this suggestion we can reformulate the principle of induction to read, 'if a large number of As have been observed under a wide variety of conditions, and if all these observed As have the property B, then all As probably have the property B'. This reformulation does not overcome the problem of induction. The reformulated principle is still a universal statement. It implies, on the basis of a finite number of successes, that all applications of the principle will lead to general conclusions that are probably true. Consequently, attempts to justify the probabilistic version of the principle of induction by an appeal to experience involve an appeal to inductive arguments of the kind that are being justified just as the principle in its original form did.

There is another basic problem with interpretations of inductive arguments that construe them as leading to probable truth rather than truth. This problem arises as soon as one tries to be precise about just how probable a law or theory is in the light of specified evidence. It may seem intuitively plausible that as the observational support for a general law increases the probability that it is true also increases. But this intuition does not stand up to inspection. Given standard probability theory, it is difficult to avoid the conclusion that the probability of any general law is zero whatever the observational evidence. To make the point in a non-technical way, any observational evidence will consist of a finite number of observation statements, whereas the general law will make claims about an infinity of possible cases. The probability of the law in the light of the evidence is thus a finite number divided by infinity, which remains zero by whatever factor the finite amount of evidence is increased. Looking at it in another way, there will always be an infinite number of general statements that are compatible with a finite number of observation statements, just as there is an infinity of curves that can be drawn through a finite number of points. That is, there will always be an infinite number of hypotheses compatible with a finite amount of evidence. Consequently, the probability of any one of them being true is zero. In chapter 12 we will discuss a possible way around this problem.

In this and the preceding section we have revealed two kinds of problem with the idea that scientific knowledge is derived from the facts by some kind of inductive inference. The first concerned the issue of specifying just what an adequate inductive argument is. The second involved the circularity involved in attempts to justify induction. I regard the former problem as more severe than the latter. The reason that I do not take the problem of induction too seriously is that any attempt to provide an account of science is bound to confront a problem of a similar kind. We are bound to run into trouble if we seek rational justifications of every principle we use, for we cannot provide a *rational argument* for rational argument itself without assuming what we are arguing for. Not even logic can be *argued for* in a way that is not question begging. However, what constitutes a valid deductive argument can be specified with a high degree of precision, whereas what constitutes a good inductive argument has not been made at all clear.

The appeal of inductivism

A concise expression of the inductivist view of science, the view that scientific knowledge is derived from the facts by inductive inference, which we have discusssed in the opening chapters of this book, is contained in the following passage written by a twentieth-century economist, A. B. Wolfe.

> If we try to imagine how a mind of superhuman power and reach, but normal so far as the logical processes of its thought are concerned . . . would use the scientific method, the process would be as follows: First, all facts would be observed and recorded, *without selection* or *a priori* guess as to their relative importance. Secondly, the observed and recorded facts would be analysed, compared and classified, without *hypothesis or postulates*, other than those necessarily involved in the logic of thought. Third, from this analysis of the facts, generalizations would be inductively drawn as to the relations, classificatory or causal, between them. Fourth, further research would be deductive as well as inductive, employing inferences from previously established generalisations.[1]

We have seen that the idea that the collection of facts can and should take place prior to the acquisition and acceptance of any knowledge does not bear analysis. To suggest otherwise is to believe that my observations of the flora in the Australian bush will be of more value than those of a trained botanist precisely because I know little botany. Let us reject this part of our economist's characterisation of science. What remains is an account that has a certain appeal. It is summarised in Figure 2. The laws and theories that make up scientific knowledge are derived by induction from a factual basis supplied by observation and experiment. Once such general knowledge is available, it can be drawn on to make predictions and offer explanations.

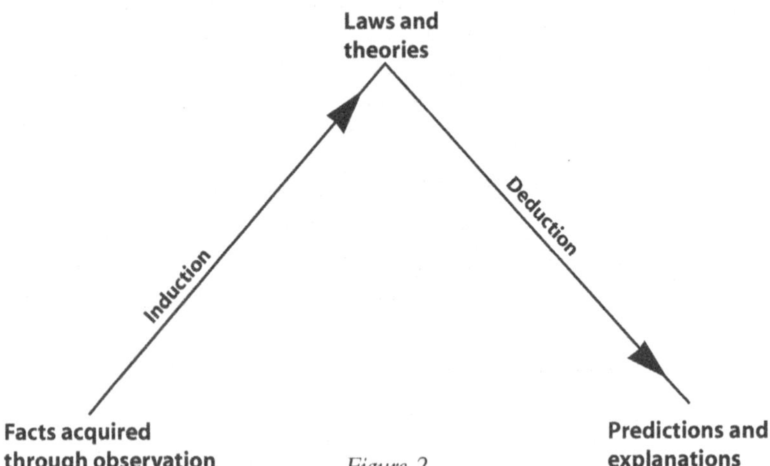

Figure 2

Consider the following argument:

1. Fairly pure water freezes at about 0°C (if given sufficient time).
2. My car radiator contains fairly pure water.

---

3. If the temperature falls well below 0°C, the water in my car radiator will freeze (if given sufficient time).

Here we have an example of a valid logical argument to deduce the prediction 3 from the scientific knowledge contained in premise 1. If 1 and 2 are true, 3 must be true. However, the truth of 1, 2 or 3 is not established by this or any other deduction. For the inductivist the source of scientific truth is experience not logic. On this view, 1 will be ascertained by direct observation of various instances of freezing water. Once 1 and 2 have been established by observation and induction, then the prediction 3 can be deduced from them.

Less trivial examples will be more complicated, but the roles played by observation, induction and deduction remain essentially the same. As a final example, I will consider the inductivist account of how physical science is able to explain the rainbow.

The simple premise 1 of the previous example is here replaced by a number of laws governing the behaviour of light, namely, the laws of reflection and refraction of light and assertions about the dependence of the amount of refraction on the colour of the light. These general laws are to be derived from experience by induction. A large number of laboratory experiments are performed, reflecting rays of light from mirrors and water surfaces, measuring angles of refraction for rays of light passing from air to water, water to air and so on, under a wide variety of circumstances, until whatever conditions are presumed to be necessary to warrant the inductive derivation of the laws of optics from the experimental results are satisfied.

Premise 2 of our previous example will also be replaced by a more complex array of statements. These will include assertions to the effect that the sun is situated in some specified position in the sky relative to an observer on earth, and that raindrops are falling from a cloud situated in some specified region relative to the observer. Sets of statements like these, which describe the set-up under investigation, will be referred to as *initial conditions*. Descriptions of experimental set-ups will be typical initial conditions.

Given the laws of optics and the initial conditions, it is now possible to perform deductions yielding an explanation of the formation of a rainbow visible to the observer. These deductions will no longer be as self-evident as in our previous examples, and will involve mathematical as well as verbal arguments. The derivation will run roughly as follows. If we assume a raindrop to be roughly spherical, then the path of a ray of light through a raindrop will be roughly as depicted in Figure 3. For a ray of white light from the sun incident

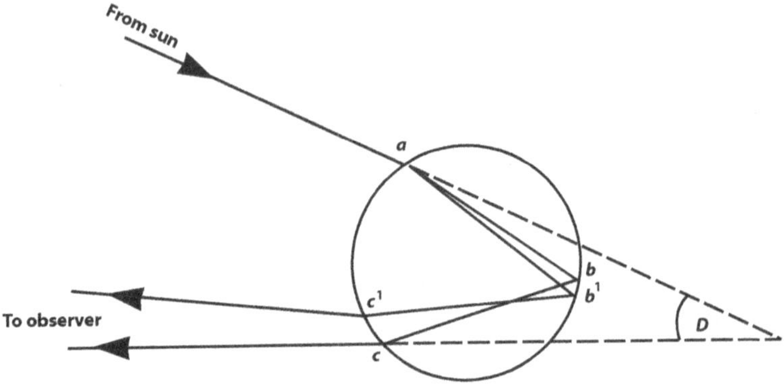

Figure 3

on a raindrop at *a*, the red light will travel along *ab* and the blue light along ab^1 according to the law of refraction. The law of reflection requires that *ab* be reflected along *bc* and ab^1 along b^1c^1. Refraction at *c* and c^1 will again be determined by the law of refraction, so that an observer viewing the raindrop will see the red and blue components of the white light separated (and also all the other colours of the spectrum). The same separation of colours will be visible to our observer for any raindrop that is situated in a region of the sky such that the line joining the raindrop to the sun makes an angle *D* with the line joining the raindrop to the observer. Geometrical considerations yield the conclusion that a coloured arc will be visible to the observer provided the rain cloud is sufficiently extended.

I have only sketched an explanation of the rainbow here, but it should suffice to illustrate the general form of the reasoning involved. Given that the laws of optics are true (and for the unqualified inductivist this can be established from observation by induction), and given that the initial conditions are correctly described, then the explanation of the rainbow necessarily follows. The general form of all scientific explanations and predictions can be summarised thus:

1. Laws and theories
2. Initial conditions

---

3. Predictions and explanations

This is the step depicted on the right-hand side of Figure 2.

The basic inductivist account of science does have some immediate appeal. Its attraction lies in the fact that it does seem to capture in a formal way some of the commonly held intuitions about the special characteristics of scientific knowledge, namely, its objectivity, its reliability and its usefulness. We have discussed the inductivist account of the usefulness of science insofar as it can facilitate predictions and explanations already in this section.

The objectivity of science as construed by the inductivist derives from the extent to which observation, induction and deduction are themselves seen as objective. Observable facts are understood to be established by an unprejudiced use of the senses in a way that leaves no room for subjective opinion to intrude. As far as inductive and deductive reasoning are concerned, these are adequate to the extent that they conform to publicly formulated criteria of adequacy, so, once again, there is no room for personal opinion. Inferences either conform to the objective standards or they don't.

The reliability of science follows from the inductivist's claims about observation and both inductive and deductive reasoning. According to the unqualified inductivist, observation statements that form the factual basis for science can be securely established directly by careful use of the senses. Further, this security will be transmitted to the laws and theories inductively derived from those facts provided the conditions for adequate inductive generalisations are met. This is guaranteed by the principle of induction which is presumed to form the basis of science.

Attractive as it may have appeared, we have seen that the inductivist position is, at best, in need of severe qualification and, at worst, thoroughly inadequate. We have seen that facts adequate for science are by no means straightforwardly given but have to be practically constructed, are in some important senses dependent on the knowledge that they presuppose, a complication overlooked in the schematisation in Figure 2, and are subject to improvement and replacement. More seriously, we have been unable to give a precise specification of induction in a way that will help distinguish a justifiable generalisation from the facts from a hasty or rash one, a formidable task given nature's capacity to surprise, epitomised in the discovery that supercooled liquids can flow uphill.

In chapter 12 we will discuss some recent attempts to rescue the inductivist account of science from its difficulties. Meanwhile, we will turn in the next two chapters to a philosopher who attempts to sidestep problems with induction by putting forward a view of science that does not involve induction.

Further reading

The historical source of Hume's problem of induction is Hume's *Treatise on Human Nature* (1939, Part 3). Another classic discussion of the problem is Russell (1912, chapter 6). A thorough, technical investigation of the consequences of Hume's argument is Stove (1973). Karl Popper's claim to have solved the problem of induction is in Popper (1979, chapter 1). Reasonably accessible accounts of inductive reasoning can be found in Hempel (1966) and Salmon (1966), and a more detailed treatment is found in Glymour (1980). See also Lakatos (1968) for a collection of essays, including a provocative survey by Lakatos himself, of attempts to construct an inductive logic. Musgrave (1993) is a good historical survey of the problem.

CHAPTER 5

Introducing falsificationism

Introduction

Karl Popper was the most forceful advocate of an alternative to inductivism that I will refer to as 'falsificationism'. Popper was educated in Vienna in the 1920s, at a time when logical positivism was being articulated by a group of philosophers who became known as the Vienna Circle. One of the most famous of these was Rudolph Carnap, and the clash and debate between his supporters and those of Popper was to be a feature of philosophy of science up until the 1960s. Popper himself tells the story of how he became disenchanted with the idea that science is special because it can be derived from the facts, the more facts the better. He became suspicious of the way in which he saw Freudians and Marxists supporting their theories by interpreting a wide range of instances, of human behaviour or historical change respectively, in terms of their theory and claiming them to be supported on this account. It seemed to Popper that these theories could never go wrong because they were sufficiently flexible to accommodate any instances of human behaviour or historical change as compatible with their theory. Consequently, although giving the appearance of being powerful theories confirmed by a wide range of facts, they could in fact explain nothing because they could rule out nothing. Popper compared this with a famous test of Einstein's theory of general relativity carried out by Eddington in 1919. Einstein's theory had the implication that rays of light should bend as they pass close to massive objects such as the sun. As a consequence, a star situated beyond the sun should appear displaced from the direction in which it would be observed in the absence of this bending. Eddington sought for this displacement by sighting the star at a time when the light from the sun was blocked out by an eclipse. It transpired that the displacement was observed and Einstein's theory was borne out. But Popper makes the point that it might not have been. By making a specific, testable prediction the general theory of relativity was at risk. It ruled out observations that clashed with that prediction. Popper drew the moral that genuine scientific theories, by making definite predictions, rule out a range of observable states of affairs in a way that he considered Freudian and Marxist theory failed to do. He arrived at his key idea that scientific theories are *falsifiable*.

Falsificationists freely admit that observation is guided by and presupposes theory. They are also happy to abandon any claim implying that theories can be established as true or probably true in the light of observational evidence. Theories are construed as speculative and tentative conjectures or guesses freely created by the human intellect in an attempt to overcome problems encountered by previous theories to give an adequate account of some aspects of the world or universe. Once proposed, speculative theories are to be rigorously and ruthlessly tested by observation and experiment. Theories that fail to stand up to observational and experimental tests must be eliminated and replaced by further speculative conjectures. Science progresses by trial and error, by conjectures and refutations. Only the fittest theories survive. Although it can never be legitimately said of a theory that it is true, it can hopefully be said that it is the best available; that it is better than anything that has come before. No problems about the charactisation and justification of induction arises for the falsificationists because, according to them, science does not involve induction.

The content of this condensed summary of falsificationism will be filled out in the next two chapters.

A logical point in favour of falsificationism

According to falsificationism, some theories can be shown to be false by an appeal to the results of observation and experiment. There is a simple, logical point that seems to support the falsificationist here. I have already indicated in chapter 4 that, even if we assume that true observational statements are available to us in some way, it is never possible to arrive at universal laws and theories by logical deductions on that basis alone. However, it is possible to perform logical deductions starting from singular observation statements as premises, to arrive at the falsity of universal laws and theories by logical deduction. For example, if we are given the statement, 'A raven which was not black was observed at place *x* at time *t*', then it logically follows from this that 'All ravens are black' is false. That is, the argument:

Premise	A raven, which was not black, was at place *x* at time *t*.
Conclusion	Not all ravens are black.

is a logically valid deduction. If the premise is asserted and the conclusion denied, a contradiction is involved. One or two more examples will help illustrate this fairly trivial logical point. If it can be established by observation in some test experiment that a ten-kilogram weight and a one-kilogram weight in free fall move downwards at roughly the same speed, then it can be concluded that the claim that bodies fall at speeds proportional to their weight is false. If it can be demonstrated beyond doubt that a ray of light passing close to the sun is deflected in a curved path, then it is not the case that light necessarily travels in straight lines.

The falsity of universal statements can be deduced from suitable singular statements. The falsificationist exploits this logical point to the full.

Falsifiability as a criterion for theories

The falsificationist sees science as a set of hypotheses that are tentatively proposed with the aim of accurately describing or accounting for the behaviour of some aspect of the world or universe. However, not any hypothesis will do. There is one fundamental condition that any hypothesis or system of hypotheses must satisfy if it is to be granted the status of a scientific law or theory. If it is to form part of science, a hypothesis must be *falsifiable*. Before proceeding any further, it is important to be clear about the falsificationist's usage of the term 'falsifiable'.

Here are some examples of some simple assertions that are falsifiable in the sense intended.

1. It never rains on Wednesdays.
2. All substances expand when heated.
3. Heavy objects such as a brick when released near the surface of the earth fall straight downwards if not impeded.
4. When a ray of light is reflected from a plane mirror, the angle of incidence is equal to the angle of reflection.

Assertion 1 is falsifiable because it can be falsified by observing rain to fall on a Wednesday. Assertion 2 is falsifiable. It can be falsified by an observation statement to the effect that some substance, x, did not expand when heated at time t. Water near its freezing point would serve to falsify 2. Both 1 and 2 are falsifiable and false. Assertions 3 and 4 may be true, for all I know. Nevertheless, they are falsifiable in the sense intended. It is logically possible that the next brick to be released will 'fall' upwards. No logical contradiction is involved in the assertion, 'The brick fell upwards when released', although it may be that no such statement is ever supported by observation. Assertion 4 is falsifiable because a ray of light incident on a mirror at some oblique angle could conceivably be reflected in a direction perpendicular to the mirror. This will never happen if the law of reflection happens to be true, but no logical contradiction would be involved if it did. Both 3 and 4 are falsifiable, even though they may be true.

A hypothesis is falsifiable if there exists a logically possible observation statement or set of observation statements that are inconsistent with it, that is, which, if established as true, would falsify the hypothesis.

Here are some examples of statements that do not satisfy this requirement and that are consequently not falsifiable.

5. Either it is raining or it is not raining.
6. All points on a Euclidean circle are equidistant from the centre.
7. Luck is possible in sporting speculation.

No logically possible observation statement could refute 5. It is true whatever the weather is like. Assertion 6 is necessarily true because of the definition of a Euclidean circle. If points on a circle were not equidistant from some fixed point, then that figure would just not be a Euclidean circle. 'All bachelors are unmarried' is unfalsifiable for a similar reason. Assertion 7 is quoted from a horoscope in a newspaper. It typifies the fortune-teller's devious strategy. The assertion is unfalsifiable. It amounts to telling the reader that if he has a bet today he might win, which remains true whether he bets or not, and if he does, whether he wins or not.

Falsificationists demand that scientific hypotheses be falsifiable, in the sense discussed. They insist on this because it is only by ruling out a set of logically possible observation statements that a law or theory is informative. If a statement is unfalsifiable, then the world can have any properties whatsoever, and can behave in any way whatsoever, without conflicting with the statement. Assertions 5, 6 and 7, unlike assertions 1, 2, 3 and 4, tell us nothing about the world. A scientific law or theory should ideally give us some information about how the world does in fact behave, thereby ruling out ways in which it could (logically) possibly behave but in fact does not. The law 'All planets move in ellipses around the sun' is scientific because it claims that planets in fact move in ellipses and rules out orbits that are square or oval. Just because the law makes definite claims about planetary orbits, it has informative content and is falsifiable.

A cursory glance at some laws that might be regarded as typical components of scientific theories indicates that they satisfy the falsifiability criterion. 'Unlike magnetic poles attract each other', 'An acid added to a base yields a salt plus water' and similar laws can easily be construed as falsifiable. However, the falsificationist maintains that some theories, while they may superficially appear to have the characteristics of good scientific theories, are in fact only posing as scientific theories because they are not falsifiable and should be rejected. Popper has claimed that some versions at least of Marx's theory of history, Freudian psychoanalysis and Adlerian psychology suffer from this fault. The point can be illustrated by the following caricature of Adlerian psychology.

A fundamental tenet of Adler's theory is that human actions are motivated by feelings of inferiority of some kind. In our caricature, this is supported by the following incident. A man is standing on the bank of a treacherous river at the instant a child falls into the river nearby. The man will either leap into the river in an attempt to save the child or he will not. If he does leap in, the Adlerian responds by indicating how this supports his theory. The man obviously needed to overcome his feelings of inferiority by demonstrating that he was brave enough to leap into the river, in spite of the danger. If the man does not leap in, the Adlerian can again claim support for his theory. The man was overcoming his feelings of inferiority by demonstrating that he had the strength of will to remain on the bank, unperturbed, while the child drowned.

If this caricature is typical of the way in which Adlerian theory operates, then the theory is not falsifiable. It is consistent with any kind of human behaviour, and just because of that, it tells us nothing about human behaviour. Of course, before Adler's theory can be rejected on these grounds, it would be necessary to investigate the details of the theory rather than a caricature. But there are plenty of social, psychological and religious theories that give rise to the suspicion that in their concern to explain everything they explain nothing. The existence of a loving God and the occurrence of some disaster can be made compatible by interpreting the disaster as being sent to try us or to punish us, whichever seems most suited to the situation. Many examples of animal behaviour can be seen as evidence supporting the assertion, 'Animals are designed so as best to fulfil the function for which they were intended'. Theorists operating in this way are guilty of the fortune-teller's evasion and are subject to the falsificationist's criticism. If a theory is to have informative content, it must run the risk of being falsified.

Degree of falsifiability, clarity and precision

A good scientific law or theory is falsifiable just because it makes definite claims about the world. For the falsificationist, it follows fairly readily from this that the more falsifiable a theory is the better, in some loose sense of more. The more a theory claims, the more potential opportunities there will be for showing that the world does not in fact behave in the way laid down by the theory. A very good theory will be one that makes very wide-ranging claims about the world, and which is consequently highly falsifiable, and is one that resists falsification whenever it is put to the test.

The point can be illustrated by means of a trivial example. Consider these laws:

(a) Mars moves in an ellipse around the sun.
(b) All planets move in ellipses around their sun.

I take it that it is clear that (b) has a higher status than (a) as a piece of scientific knowledge. Law (b) tells us all that (a) tells us and more besides. Law (b), the preferable law, is more falsifiable than (a). If observations of Mars should turn out to falsify (a), then they would falsify (b) also. Any falsification of (a) will be a falsification of (b), but the reverse is not the case. Observation statements referring to the orbits of Venus, Jupiter, etc. that might conceivably falsify (b) are irrelevant to (a). If we follow Popper and refer to those sets of observation statements that would serve to falsify a law or theory as *potential falsifiers* of that law or theory, then we can say that the potential falsifiers of (a) form a class that is a subclass of the potential falsifiers of (b). Law (b) is more falsifiable than law (a), which is tantamount to saying that it claims more, that it is the better law.

A less-contrived example involves the relation between Kepler's theory of the solar system and Newton's. Kepler's theory I take to be his three laws of planetary motion. Potential falsifiers of that theory consist of sets of statements referring to planetary positions relative to the sun at specified times. Newton's theory, a better theory that superseded Kepler's, is more comprehensive. It consists of Newton's laws of motion plus his law of gravitation, the latter asserting that all pairs of bodies in the universe attract each other with a force that varies inversely as the square of their separation. Some of the potential falsifiers of Newton's theory are sets of statements of planetary positions at specified times. But there are many others, including those referring to the behaviour of falling bodies and pendulums, the correlation between the tides and the locations of the sun and moon, and so on. There are many more opportunities for falsifying Newton's theory than for falsifying Kepler's theory. And yet, so the falsificationist story goes,

Newton's theory was able to resist attempted falsifications, thereby establishing its superiority over Kepler's.

Highly falsifiable theories should be preferred to less falsifiable ones, then, provided they have not in fact been falsified. The qualification is important for the falsificationist. Theories that have been falsified must be ruthlessly rejected. The enterprise of science involves the proposal of highly falsifiable hypotheses, followed by deliberate and tenacious attempts to falsify them. To quote Popper (1969, p. 231, italics in original):

> I can therefore gladly admit that falsificationists like myself much prefer an attempt to solve an interesting problem by a bold conjecture, *even (and especially) if it soon turns out to be false*, to any recital of a sequence of irrelevant truisms. We prefer this because we believe that this is the way in which we can learn from our mistakes, and that in finding that our conjecture was false we shall have learnt much about the truth, and shall have got nearer to the truth.

We learn from our *mistakes*. Science progresses by trial and *error*. Because of the logical situation that renders the derivation of universal laws and theories from observation statements impossible, but the deduction of their falsity possible, *falsifications* become the important landmarks, the striking achievements, the major growing-points in science. This somewhat counter-intuitive emphasis of the more extreme falsificationists on the significance of falsifications will be criticised in later chapters.

Because science aims at theories with a large informative content, the falsificationist welcomes the proposal of bold speculative conjectures. Rash speculations are to be encouraged, provided they are falsifiable and provided they are rejected when falsified. This do-or-die attitude clashes with the caution advocated by the extreme inductivist. According to the latter, only those theories that can be shown to be true or probably true are to be admitted into science. We should proceed beyond the immediate results of experience only so far as legitimate inductions will take us. The falsificationist, by contrast, recognises the limitation of induction and the subservience of observation to theory. Nature's secrets can only be revealed with the aid of ingenious and penetrating theories. The greater the number of conjectured theories that are confronted by the realities of the world, and the more speculative those conjectures are, the greater will be the chances of major advances in science. There is no danger in the proliferation of speculative theories because any that are inadequate as descriptions of the world can be ruthlessly eliminated as the result of observational or other tests.

The demand that theories should be highly falsifiable has the attractive consequence that theories should be clearly stated and precise. If a theory is so vaguely stated that it is not clear exactly what it is claiming, then when tested by observation or experiment it can always be interpreted so as to be consistent with the results of those tests. In this way, it can be defended against falsifications. For example, Goethe (1970, p. 295) wrote of electricity that:

> it is nothing, a zero, a mere point, which, however, dwells in all apparent existences, and at the same time is the point of origin whence, on the slightest stimulus, a double appearance presents itself, an appearance which only manifests itself to vanish. The conditions under which this manifestation is excited are infinitely varied, according to the nature of particular bodies.

If we take this quotation at face value, it is very difficult to see what possible set of physical circumstances could serve to falsify it. Just because it is so vague and indefinite (at least when taken out of context), it is unfalsifiable. Politicians and fortune-tellers can avoid being accused of making mistakes by making their assertions so vague that they can always be construed as compatible with whatever may eventuate. The demand for a high degree of falsifiability rules out such manoeuvres. The falsificationist demands that theories be stated with sufficient clarity to run the risk of falsification.

A similar situation exists with respect to precision. The more precisely a theory is formulated the more falsifiable it becomes. If we accept that the more falsifiable a theory is the better (provided it has not been falsified), then we must also accept that the more precise the claims of a theory are the better. 'Planets move in ellipses around the sun' is more precise than 'Planets move in closed loops around the sun', and is consequently more falsifiable. An oval orbit would falsify the first but not the second, whereas an orbit that falsifies the second will also falsify the first. The falsificationist is committed to preferring the first. Similarly, the falsificationist must prefer the claim that the velocity of light in a vacuum is 299.8×10^6 metres per second to the less-precise claim that it is about 300×10^6 metres per second, just because the first is more falsifiable than the second.

The closely associated demands for precision and clarity of expression both follow naturally from the falsificationist's account of science.

Falsificationism and progress

The progress of science as the falsificationist sees it might be summed up as follows. Science starts with problems, problems associated with the explanation of the behaviour of some aspects of the world or universe. Falsifiable hypotheses are proposed by scientists as solutions to a problem. The conjectured hypotheses are then criticised and tested. Some will be quickly eliminated. Others might prove more successful. These must be subject to even more stringent criticism and testing. When a hypothesis that has successfully withstood a wide range of rigorous tests is eventually falsified, a new problem, hopefully far removed from the original solved problem, has emerged. This new problem calls for the invention of new hypotheses, followed by renewed criticism and testing. And so the process continues indefinitely. It can never be said of a theory that it is true, however well it has withstood rigorous tests, but it can hopefully be said that a current theory is superior to its predecessors in the sense that it is able to withstand tests that falsified those predecessors.

Before we look at some examples to illustrate this falsificationist conception of the progress of science, a word should be said about the claim that 'Science starts with problems'. Here are some problems that have confronted scientists in the past. How are bats able to fly so dexterously at night, when in fact they have very small, weak eyes? Why is the height of a simple barometer lower at high altitudes than at low altitudes? Why were the photographic plates in Roentgen's laboratory continually becoming blackened? Why does the perihelion of the planet Mercury advance? These problems arise from more or less straightforward *observations*. In insisting on the fact that science starts with problems, then, is it not the case that, for the falsificationist just as for the naive inductivist, science starts from observation? The answer to this question is a firm 'No'. The observations cited above as constituting problems are only problematic *in the light of some theory*. The first is problematic in the light of the theory that living organisms 'see' with their eyes; the second was problematic for the supporters of Galileo's theories because it clashed with the 'force of a vacuum' theory accepted by them as an explanation of why the mercury does not fall from a barometer tube; the third was problematic for Roentgen because it was tacitly assumed at the time that no radiation or emanation of any kind existed that could penetrate the container of the photographic plates and darken them; the fourth was problematic because it was incompatible with Newton's theory. The claim that science starts with problems is perfectly compatible with the priority of theories over observation and observation statements. Science does not start with stark observation.

After this digression, we return to the falsificationist conception of the progress of science as the progression from problems to speculative hypotheses, to their criticism and eventual falsification and thence to new problems. Two examples will be offered, the first a simple one concerning the flight of bats, the second a more ambitious one concerning the progress of physics.

We start with a problem. Bats are able to fly with ease and at speed, avoiding the branches of trees, telegraph wires, other bats, etc., and can catch insects. And yet bats have weak eyes, and in any case do most of their flying at night. This poses a problem because it apparently falsifies the plausible theory that animals, like humans, see with their eyes. A falsificationist will attempt to solve the problem by making a conjecture or hypothesis. Perhaps he suggests that, although bats' eyes are apparently weak, nevertheless in some way that is not understood they are able to see efficiently at night by use of their eyes. This hypothesis can be tested. A sample of bats is released into a darkened room containing obstacles and their ability to avoid the obstacles measured in some way. The same bats are now blindfolded and again released into the room. Prior to the experiment, the experimenter can make the following deduction. One premise of the deduction is his hypothesis, which made quite explicit reads, 'Bats are able to fly avoiding obstacles by using their eyes, and cannot do so without the use of their eyes'. The second premise is a description of the experimental set-up, including the statement, 'This sample of bats is blindfolded so that they do not have the use of their eyes'. From these two premises, the experimenter can derive, deductively, that the sample of bats will not be able to avoid the obstacles in the test laboratory efficiently. The experiment is now performed and it is found that the bats avoid collisions just as efficiently as before. The hypothesis has been falsified. There is now a need for a fresh use of the imagination, a new conjecture or hypothesis or guess. Perhaps a scientist suggests that in some way the bat's ears are involved in its ability to avoid obstacles. The hypothesis can be tested, in an attempt to falsify it, by plugging the ears of bats before releasing them into the test laboratory. This time it is found that the ability of the bats to avoid obstacles is considerably impaired. The hypothesis has been supported. The falsificationist must now try to make the hypothesis more precise so that it becomes more readily falsifiable. It is suggested that the bat hears echoes of its own squeaks rebounding from solid objects. This is tested by gagging the bats before releasing them. Again the bats collide with obstacles and again the hypothesis is supported. The falsificationist now appears to be reaching a tentative solution to the problem, although it has not been *proved* by experiment how bats avoid collisions while flying. Any number of factors may turn up that show the hypothesis to have been wrong. Perhaps the bat detects echoes not with its ears but with sensitive regions close to the ears, the functioning of which was impaired when the bat's ears were plugged. Or perhaps different kinds of bats detect obstacles in very different ways, so the bats used in the experiment were not truly representative.

The progress of physics from Aristotle through Newton to Einstein provides an example on a larger scale. The falsificiationist account of that progression goes something like this. Aristotelian physics was to some extent quite successful. It could explain a wide range of phenomena. It could explain why heavy objects fall to the ground (seeking their natural place at the centre of the universe), it could explain the action of siphons and liftpumps (the explanation being based on the impossibility of a vacuum), and so on. But eventually Aristotelian physics was falsified in a number of ways. Stones dropped from the top of the mast of a uniformly moving ship fell to the deck at the foot of the mast and not some distance from the mast, as Aristotle's theory predicted. The moons of Jupiter can be seen to orbit Jupiter and not the earth. A host of other falsifications were accumulated during the seventeenth century. Newton's physics, however, once it had been created and developed by way of the conjectures of the likes of Galileo and Newton, was a superior theory that superseded Aristotle's. Newton's theory could account for falling objects, the operation of siphons and liftpumps and anything else that Aristotle's theory could explain, and could also account for the phenomena that were problematic for the Aristotelians. In addition, Newton's theory could explain phenomena not touched on by Aristotle's theory, such as correlations between the tides and the location of the moon, and the variation in the force of gravity with height above sea level. For two centuries Newton's theory was successful. That is, attempts to falsify it by reference to the new phenomena predicted with its help were unsuccessful. The theory even led to the discovery of a new planet, Neptune. But in spite of its success, sustained attempts to falsify it eventually proved successful. Newton's theory was falsified in a number of ways. It was unable to account for the details of the orbit of the planet Mercury and was unable to account for the variable mass of fast-moving electrons in discharge tubes. Challenging problems faced physicists, then, as the nineteenth century gave way to the twentieth, problems calling for new speculative hypotheses designed to overcome these problems in a progressive way. Einstein was able to meet this challenge. His relativity theory was able to account for the phenomena that falsified Newton's theory, while at the same time being able to match Newton's theory in those areas where the latter had proved successful. In addition, Einstein's theory yielded the prediction of spectacular new phenomena. His special theory of relativity predicted that mass should be a function of velocity and that mass and energy could be transformed into one another, and his general theory predicted that light rays should be bent by strong gravitational fields. Attempts to refute Einstein's theory by reference to the new phenomena failed. The falsification of Einstein's theory remains a challenge for modern physicists. Their success, if it should eventuate, would mark a new step forward in the progress of physics.

So runs a typical falsification account of the progress of physics. Later we shall have cause to doubt its accuracy and validity.

From the foregoing, it is clear that the concept of progress, of the growth of science, is a conception that is a central one in the falsificationist account of science. This issue is pursued in more detail in the next chapter.

Further reading

The classic falsificationist text is Popper in *The Logic of Scientific Discovery* (1972), first published in German in 1934 and translated into English in 1959. More recent collections of his writings are Popper (1969) and Popper (1979). Popper's own story about how he came to his basic idea through comparing Freud, Adler and Marx with Einstein is in chapter 1 of his 1969 text. More sources related to falsificationism will be given at the end of the next chapter.

CHAPTER 6

Sophisticated falsificationism, novel predictions and the growth of science

Relative rather than absolute degrees of falsifiability

The previous chapter mentioned some conditions that a hypothesis should satisfy in order to be worthy of a scientist's consideration. A hypothesis should be falsifiable, the more falsifiable the better, and yet should not be falsified. More sophisticated falsificationists realise that those conditions alone are insufficient. A further condition is connected with the need for science to progress. A hypothesis should be more falsifiable than the one for which it is offered as a replacement.

The sophisticated falsificationist account of science, with its emphasis on the growth of science, switches the focus of attention from the merits of a single theory to the relative merits of competing theories. It gives a dynamic picture of science rather than the static account of the most naive falsificationists. Instead of asking of a theory, 'Is it falsifiable?', 'How falsifiable is it?' and 'Has it been falsified?', it becomes more appropriate to ask, 'Is this newly proposed theory a viable replacement for the one it challenges?' In general, a newly proposed theory will be acceptable as worthy of the consideration of scientists if it is more falsifiable than its rival, and especially if it predicts a new kind of phenomenon not touched on by its rival.

The emphasis on the comparison of degrees of falsifiability of series of theories, which is a consequence of the emphasis on a science as a growing and evolving body of knowledge, enables a technical problem to be bypassed. For it is very difficult to specify just how falsifiable a single theory is. An absolute measure of falsifiability cannot be defined simply because the number of potential falsifiers of a theory will always be infinite. It is difficult to see how the question 'How falsifiable is Newton's law of gravitation?' could be answered. On the other hand, it is often possible to compare the degrees of falsifiability of laws or theories. For instance, the claim 'All pairs of bodies attract each other with a force that varies inversely as the square of their separation' is more falsifiable than the claim. 'The planets in the solar system attract each other with a force that varies inversely as the square of their separation'. The second is implied by the first. Anything that falsifies the second will falsify the first, but the reverse is not true. Ideally, the falsificationist would like to be able to say that the series of theories that constitute the historical evolution of a science is made up of falsifiable theories, each one in the series being more falsifiable than its predecessor.

Increasing falsifiability and ad hoc modifications

The demand that as a science progresses its theories should become more and more falsifiable, and consequently have more and more content and be more and more informative, rules out modifications in theories that are designed merely to protect a theory from a threatening falsification. A modification in a theory, such as the addition of an extra postulate or a change in some existing postulate, that has no testable consequences that were not already testable consequences of the unmodified theory will be called *ad hoc* modifications. The remainder of this section will consist of examples designed to clarify the notion of an ad hoc modification. I will first consider some ad hoc modifications, which the falsificationist would reject, and afterwards these will be contrasted with some modifications that are not ad hoc and which the falsificationist would consequently welcome.

I begin with a rather trivial example. Let us consider the generalisation 'Bread nourishes'. This low-level theory, if spelt out in more detail, amounts to the claim that if wheat is grown in the normal way, converted into bread in the normal way and eaten by humans in a normal way, then those humans will be nourished. This apparently innocuous theory ran into trouble in a French village on an occasion when wheat was grown in a normal way, converted into bread in a normal way and yet most people who ate the bread became seriously ill and many died. The theory '(All) bread nourishes' was falsified. The theory can be modified to avoid this falsification by adjusting it to read, '(All) bread, with the exception of that particular batch of bread produced in the French village in question, nourishes'. This is an ad hoc modification. The modified theory cannot be tested in any way that was not also a test of the original theory. The consuming of any bread by any human constitutes a test of the original theory, whereas tests of the modified theory are restricted to the consuming of bread other than that batch of bread that led to such disastrous results in France. The modified hypothesis is less falsifiable than the original version. The falsificationist rejects such rearguard actions.

The next example is less gruesome and more entertaining. It is an example based on an interchange that actually took place in the seventeenth century between Galileo and an Aristotelian adversary. Having carefully observed the moon through his newly invented telescope, Galileo was able to report that the moon was not a smooth sphere but that its surface abounded in mountains and craters. His Aristotelian adversary had to admit that things did appear that way when he repeated the observations for himself. But the observations threatened a notion fundamental for many Aristotelians, namely, that all celestial bodies are perfect spheres. Galileo's rival defended his theory in the face of the apparent falsification in a way that was blatantly ad hoc. He suggested that there was an invisible substance on the moon filling the craters and covering the mountains in such a way that the moon's shape was perfectly spherical. When Galileo inquired how the presence of the invisible substance might be detected, the reply was that there was no way in which it could be detected. There is no doubt, then, that the modified theory led to no new testable consequences and would be quite unacceptable to a falsificationist. An exasperated Galileo was able to show up the inadequacy of his rival's position in a characteristically witty way. He announced that he was prepared to admit that the invisible, undetectable substance existed on the moon, but insisted that it was not distributed in the way suggested by his rival but in fact was piled up on top of the mountains so that they were many times higher than they appeared through the telescope. Galileo was able to outmanoeuvre his rival in the fruitless game of the invention of ad hoc devices for the protection of theories.

One other example of a possibly ad hoc hypothesis from the history of science will be briefly mentioned. Prior to Lavoisier, the phlogiston theory was the standard theory of combustion. According to that theory, phlogiston is emitted from substances when they are burnt. This theory was threatened when it was discovered that many substances gain weight after combustion. One way of overcoming the apparent falsification was to suggest that phlogiston has negative weight. If this hypothesis could be tested only by weighing substances before and after combustion, then it was ad hoc. It led to no new tests.

Modifications of a theory in an attempt to overcome a difficulty need not be ad hoc. Here are some examples of modifications that are not ad hoc, and which consequently are acceptable from a falsificationist point of view.

Let us return to the falsification of the claim 'Bread nourishes' to see how this could be modified in an acceptable way. An acceptable move would be to replace the original falsified theory by the claim 'All bread nourishes except bread made from wheat contaminated by a particular kind of fungus' (followed by a specification of the fungus and some of its characteristics). This modified theory is not ad hoc because it leads to new tests. It is *independently testable*, to use Popper's (1972, p. 193) phrase. Possible tests would include testing the wheat from which the poisonous bread was made for the presence of the fungus, cultivating the fungus on some specially prepared wheat and testing the nourishing effect of the bread produced from it, chemically analysing the fungus for the presence of known poisons, and so on. All these tests, many of which do not constitute tests of the original hypothesis, could result in the falsification of the modified hypothesis. If the modified, more falsifiable, hypothesis resists falsification in the face of the new tests, then something new will have been learnt and progress will have been made.

Turning now to the history of science for a less artificial example, we might consider the chain of events that led to the discovery of the planet Neptune. Nineteenth-century observations of the motion of the planet Uranus indicated that its orbit departed considerably from that predicted on the basis of Newton's gravitational theory, thus posing a problem for that theory. In an attempt to overcome the difficulty, it was suggested by Leverrier in France and by Adams in England that there existed a previously undetected planet in the vicinity of Uranus. The attraction between the conjectured planet and Uranus was to account for the latter's departure from its initially predicted orbit. This suggestion was not ad hoc, as events were to show. It was possible to estimate the approximate vicinity of the conjectured planet if it were to be of a reasonable size and to be responsible for the perturbation of Uranus's orbit. Once this had been done, it was possible to test the new proposal by inspecting the appropriate region of the sky through a telescope. It was in this way that Galle came to make the first sighting of the planet now known as Neptune. Far from being ad hoc, the move to save Newton's theory from falsification by Uranus's orbit led to a new kind of test of that theory, which it was able to pass in a dramatic and progressive way.

Confirmation in the falsificationist account of science

When falsificationism was introduced as an alternative to inductivism in the previous chapter, falsifications (that is, the failures of theories to stand up to observational and experimental tests) were portrayed as being of key importance. It was argued that the logical situation permits the establishment of the falsity but not of the truth of theories in the light of available observation statements. It was also urged that science should progress by the proposal of bold, highly falsifiable conjectures as attempts to solve problems, followed by ruthless attempts to falsify the new proposals. Along with this came the suggestion that significant advances in science come about when those bold conjectures are falsified. The self-avowed falsificationist Popper says as much in the passage quoted on p. 62, where the italics are his. However, exclusive attention to falsifying instances amounts to a misrepresentation of the more sophisticated falsificationist's position. More than a hint of this is contained in the example with which the previous section concluded. The independently testable attempt to save

Newton's theory by a speculative hypothesis was a success because that hypothesis was confirmed by the discovery of Neptune and not because it was falsified.

It is a mistake to regard the falsification of bold, highly falsifiable conjectures as the occasions of significant advance in science, and Popper needs to be corrected on this point. This becomes clear when we consider the various extreme possibilities. At one extreme we have theories that take the form of bold, risky conjectures, while at the other we have theories that are cautious conjectures, making claims that seem to involve no significant risks. If either kind of conjecture fails an observational or experimental test it will be falsified, and if it passes such a test we will say it is *confirmed*. Significant advances will be marked by the *confirmation* of *bold* conjectures or the *falsification* of *cautious* conjectures. Cases of the former kind will be informative, and constitute an important contribution to scientific knowledge, simply because they mark the discovery of something that was previously unheard of or considered unlikely. The discovery of Neptune and of radio waves and Eddington's confirmation of Einstein's risky prediction that light rays should bend in strong gravitational fields all constituted significant advances of this kind. Risky predictions were confirmed. The falsification of cautious conjectures is informative because it establishes that what was regarded as unproblematically true is in fact false. Russell's demonstration that naive set theory, which was based on what appear to be almost self-evident propositions, is inconsistent is an example of an informative falsification of a conjecture apparently free from risk. By contrast, little is learnt from the *falsification* of a *bold* conjecture or the *confirmation* of a *cautious* conjecture. If a bold conjecture is falsified, then all that is learnt is that yet another crazy idea has been proved wrong. The falsification of Kepler's speculation that the spacing of the planetary orbits could be explained by reference to Plato's five regular solids does not mark one of the significant landmarks in the progress of physics. Similarly, the confirmation of cautious hypotheses is uninformative. Such confirmations merely indicate that some theory that was well established and regarded as unproblematic has been successfully applied once again. For instance, the confirmation of the conjecture that samples of iron extracted from its ore by some new process will, like other iron, expand when heated would be of little consequence.

The falsificationist wishes to reject ad hoc hypotheses and to encourage the proposal of bold hypotheses as potential improvements on falsified theories. Those bold hypotheses will lead to novel, testable predictions, which do not follow from the original, falsified theory. However, although the fact that it does lead to the possibility of new tests makes a hypothesis worthy of investigation, it will not rank as an improvement on the problematic theory it is designed to replace until it has survived at least some of those tests. This is tantamount to saying that before it can be regarded as an adequate replacement for a falsified theory, a newly and boldly proposed theory must make some novel predictions that are confirmed. Many wild and rash speculations will not survive subsequent testing and consequently will not be rated as contributing to the growth of scientific knowledge. The occasional wild and rash speculation that does lead to a novel, unlikely prediction, which is nevertheless confirmed by observation or experiment, will thereby become established as a highlight in the history of the growth of science. The *confirmations* of novel predictions resulting from bold conjectures are very important in the falsificationist account of the growth of science.

Boldness, novelty and background knowledge

A little more needs to be said about the adjectives 'bold' and 'novel' as applied to hypotheses and predictions respectively. They are both historically relative notions. What rates as a bold conjecture at one stage in the history of science may no longer be bold at some later stage. When Maxwell proposed his 'dynamical theory of the electromagnetic field' in 1864, it was a bold conjecture. It was bold because it conflicted with theories generally accepted at the time, theories that included the assumption that electromagnetic systems (magnets, charged bodies, current-carrying conductors) act upon each other instantaneously across empty space and that electromagnetic effects can be propagated at a finite velocity only through material substances. Maxwell's theory clashed with these generally accepted assumptions because it predicted that light is an electromagnetic phenomenon and also predicted, as was to be realised later, that fluctuating currents should emit a new kind of radiation, radio waves, travelling at a finite velocity through empty space. In 1864, therefore, Maxwell's theory was bold and the subsequent prediction of radio waves was a *novel* prediction. Today, the fact that Maxwell's theory can give an accurate account of the behaviour of a wide range of electromagnetic systems is a generally accepted part of scientific knowledge, and assertions about the existence and properties of radio waves will not rate as novel predictions.

If we call the complex of scientific theories generally accepted and well established at some stage in the history of science the *background knowledge* of the time, then we can say that a conjecture will be bold if its claims are unlikely in the light of the background knowledge of the time. Einstein's general theory of relativity was a bold one in 1915 because at that time background knowledge included the assumption that light travels in straight lines. This clashed with one consequence of general relativity, namely, that light rays should bend in strong gravitational fields. Copernicus's astronomy was bold in 1543 because it clashed with the background assumption that the earth is stationary at the centre of the universe. It would not be considered bold today.

Just as conjectures will be considered bold or otherwise by reference to the relevant background knowledge, so predictions will be judged novel if they involve some phenomenon that does not figure in, or is perhaps explicitly ruled out by, the background knowledge of the time. The prediction of Neptune in 1846 was a novel one because the background knowledge at that time contained no reference to such a planet. The prediction that Poisson deduced from Fresnel's wave theory of light in 1818, namely, that a bright spot should be observed at the centre of one side of an opaque disc suitably illuminated from the other, was novel because the existence of that bright spot was ruled out by the particle theory of light that formed part of the background knowledge of the time.

In the previous section it was argued that major contributions to the growth of scientific knowledge come about either when a bold conjecture is confirmed or when a cautious conjecture is falsified. The idea of background knowledge enables us to see that these two possibilities will occur together as the result of a single experiment. Background knowledge consists of cautious hypotheses just because that knowledge is well established and considered to be unproblematic. The confirmation of a bold conjecture will involve the falsification of some part of the background knowledge with respect to which the conjecture was bold.

Comparison of the inductivist and falsificationist view of confirmation

We have seen that confirmation has an important role to play in science as interpreted by the sophisticated falsificationist. However, this does not totally invalidate the labelling of that position 'falsificationism'. It is still maintained by the sophisticated falsificationist that theories can be falsified and rejected, while it is denied that theories can ever be established as true or probably true. The aim of science is to falsify theories and to replace them by better theories, theories that demonstrate a greater ability to withstand tests. Confirmations of new theories are important insofar as they constitute evidence that a new theory is an improvement on the theory it replaces, the theory that is falsified by the evidence unearthed with the aid of, and confirming, the new theory. Once a newly proposed bold theory has succeeded in ousting its rival, then it in turn becomes a new target at which stringent tests should be directed, tests devised with the aid of further boldly conjectured theories.

Because of the falsificationists' emphasis on the growth of science, their account of confirmation is significantly different from that of the inductivists. The significance of some confirming instances of a theory according to the extreme inductivist position described in chapter 4 is determined solely by the logical relationship between the observation statements that are confirmed and the theory that they support. The degree of support given to Newton's theory by Galle's observation of Neptune is no different from the degree of support given by a modern observation of Neptune. The historical context in which the evidence is acquired is irrelevant. Confirming instances are such if they give inductive support to a theory, and the greater the number of confirming instances established, the greater the support for the theory and the more likely it is to be true. This ahistorical theory of confirmation would seem to have the unappealing consequence that innumerable observations made on falling stones, planetary positions, etc., will constitute worthwhile scientific activity insofar as they will lead to increases in the estimate of the probability of the truth of the law of gravitation.

By contrast, in the falsificationist account, the significance of confirmations depends very much on their historical context. A confirmation will confer some high degree of merit on a theory if that confirmation resulted from the testing of a novel prediction. That is, a confirmation will be significant if it is estimated that it is unlikely to eventuate in the light of the background knowledge of the time. Confirmations that are foregone conclusions are insignificant. If today I confirm Newton's theory by dropping a stone to the ground, I contribute nothing of value to science. On the other hand, if tomorrow I confirm a speculative theory implying that the gravitational attraction between two bodies depends on their temperature, falsifying Newton's theory in the process, I would have made a significant contribution to scientific knowledge. Newton's theory of gravitation and some of its limitations are part of current background knowledge, whereas a temperature dependence of gravitational attraction is not. Here is one further example in support of the historical perspective that the falsificationists introduce into confirmation. Hertz confirmed Maxwell's theory when he detected the first radio waves. I also confirm Maxwell's theory whenever I listen to my radio. The logical situation is similar in the two cases. In each case, the theory predicts that radio waves should be detected and, in each case, their successful detection lends some inductive support to the theory. Nevertheless, Hertz is justly famous for the confirmation he achieved, whereas my frequent confirmations are rightly ignored in a scientific context. Hertz made a significant step forward. When I listen to my radio I am only marking time. The historical context makes all the difference.

Advantages of falsificationism over inductivism

With a summary of the basic features of falsificationism behind us, it is time to survey some of the advantages that this position can be said to have over the inductivist position, according to which scientific knowledge is inductively derived from given facts, which we discussed in earlier chapters.

We have seen that some facts, and especially experimental results, are in an important sense theory-dependent and fallible. This undermines those inductivists who require science to have an unproblematic and given factual foundation. The falsificationist recognises that facts as well as theories are fallible. Nevertheless, for the falsificationist there is an important set of facts that constitutes the testing ground for scientific theories. It consists of those factual claims that have survived severe tests. This does have the consequence that the factual basis for science is fallible, but this does not pose as big a problem for falsificationists as it does for inductivists, since the falsificationist seeks only constant improvement in science rather than demonstrations of truth or probable truth.

The inductivist had trouble specifying the criteria for a good inductive inference, and so had difficulty answering questions concerning the circumstances under which facts can be said to give significant support to theories. The falsificationist fares better in this respect. Facts give significant support to theories when they constitute severe tests of that theory. The confirmations of novel predictions are important members of this category. This helps to explain why repetition of experiments does not result in a significant increase in the empirical support for a theory, a fact that the extreme inductivist has difficulty accommodating. The conduct of a particular experiment might well constitute a severe test of a theory. However, if the experiment has been adequately performed and the theory has survived the test, then subsequent repetitions of that same experiment will not be considered as severe a test of the theory, and so will become increasingly less able to offer significant support for it. Again, whereas the inductivist has problems explaining how knowledge of the unobservable can ever be derived from observable facts, the falsificationist has no such problem. Claims about the unobservable can be severely tested, and hence supported, by exploring their novel consequences.

We have seen that inductivists have trouble characterising and justifying the inductive inferences that are meant to show theories to be true or probably true. The falsificationist claims to bypass these problems by insisting that science does not involve induction. Deduction is used to reveal the consequences of theories so that they can be tested, and perhaps falsified. But no claims are made to the effect that the survival of tests shows a theory to be true or probably true. At best, the results of such tests show a theory to be an improvement on its predecessor. The falsificationist settles for progress rather than truth.

Further reading

For Popper's mature reflections on his falsificationism see his 1983 text, *Realism and the Aim of Science.* Schilpp (1974), in the Library of Living Philosophers series, contains Popper's autobiography, a number of articles on his philosophy by critics, and Popper's reply to those critics, as well as a detailed bibliography of Popper's writings. Accessible overviews of Popper's views are Ackermann (1976) and O'Hear (1980). The modification of Popper's views involved in the section 'Confirmation in the falsificationist account of science' is discussed in more detail in Chalmers (1973).

CHAPTER 7

The limitations of falsificationism

Problems stemming from the logical situation

The generalisations that constitute scientific laws can never be logically deduced from a finite set of observable facts, whereas the falsity of a law can be logically deduced from a single observable fact with which it clashes. Establishing by observation that there is just one black swan falsifies 'all swans are white'. This is an unexceptional and undeniable point. However, using it as grounds to support a falsificationist philosophy of science is not as straightforward as it might seem. Problems emerge as soon as we progress beyond extremely simple examples, such as the one concerning the colour of swans, to more complicated cases that are closer to the kind of situation typically met with in science.

If the truth of some observation statement, O, is given, *then* the falsity of a theory T which logically entails that O is not the case can be deduced. However, it is the falsificationists themselves who insist that the observation statements that constitute the basis of science are theory-dependent and fallible. Consequently, a clash between T and O does not have the consequence that T is false. All that logically follows from the fact that T entails a prediction inconsistent with O is that either T or O is false, but logic alone cannot tell us which. When observation and experiment provide evidence that conflicts with the predictions of some law or theory, it may be the evidence that is at fault rather than the law or theory. Nothing in the logic of the situation requires that it is always the law or theory that should be rejected on the occasion of a clash with observation or experiment. A fallible observation statement might be rejected and the fallible theory with which it clashes retained. This is precisely what was involved when Copernicus's theory was retained and the naked-eye observations of the sizes of Venus and Mars, which were logically inconsistent with that theory, discarded. It is also what is involved when modern specifications of the moon's trajectory are retained and estimates of its size based on unaided observation rejected. However securely based on observation or experiment a factual claim might be, the falsificationist's position makes it impossible to rule out the possibility that advances in scientific knowledge might reveal inadequacies in that claim. Consequently, straightforward, conclusive falsifications of theories by observation are not achievable.

The logical problems for falsification do not end here. 'All swans are white' is certainly falsified if an instance of a non-white swan can be established. But simplified illustrations of the logic of a falsification such as this disguise a serious difficulty for falsificationism that arises from the complexity of any realistic test situation. A realistic scientific theory will consist of a complex of universal statements rather than a single statement like 'All swans are white'. Further, if a theory is to be experimentally tested, then more will be involved than those statements that constitute the theory under test. The theory will need to be augmented by auxiliary assumptions, such as laws and theories governing the use of any instruments used, for instance. In addition, in order to deduce some prediction the validity of which is to be experimentally tested, it will be necessary to add initial conditions such as a description of the experimental set-up. For instance, suppose an astronomical theory is to be tested by observing the position of some planet through a telescope. The theory must predict the orientation of the telescope necessary for a sighting of the planet at some specified time. The premises from which the prediction is derived will include the interconnected statements that constitute the theory under test, initial conditions such as previous positions of the planet and sun, auxiliary assumptions such as those enabling corrections to be made for refraction of light from the planet in the earth's atmosphere, and so on. Now if the prediction that follows from this maze of premises turns out to be false (in our example, if the planet does not appear at the predicted location), then all that the logic of the situation permits us to conclude is that at least one of the premises must be false. It does not enable us to identify the faulty premise. It may be the theory under test that is at fault, but alternatively it may be an auxiliary assumption or some part of the description of the initial conditions that is responsible for the incorrect prediction. A theory cannot be conclusively falsified, because the possibility cannot be ruled out that some part of the complex test situation, other than the theory under test, is responsible for an erroneous prediction. This difficulty often goes under the name of the Duhem/Quine thesis, after Pierre Duhem (1962, pp. 183–8) who first raised it and William V.O. Quine (1961) who revived it.

Here are some examples from the history of astronomy that illustrate the point.

In an example used previously, we discussed how Newton's theory was apparently refuted by the orbit of the planet Uranus. In this case, it turned out not to be the theory that was at fault but the description of the initial conditions, which did not include a consideration of the yet-to-be-discovered planet Neptune. A second example involves an argument by means of which the Danish astronomer Tycho Brahé claimed to have refuted the Copernican theory a few decades after its first publication. If the earth orbits the sun, Brahé argued, then the direction in which a fixed star is observed from earth should vary during the course of the year as the earth moves from one side of the sun to the other. But when Brahé tried to detect this predicted parallax with his instruments, which were the most accurate and sensitive ones in existence at the time, he failed. This led Brahé to conclude that the Copernican theory was false. With hindsight, it can be appreciated that it was not the Copernican theory that was responsible for the faulty prediction, but one of Brahé's auxiliary assumptions. Brahé's estimate of the distance of the fixed stars was many times too small. When his estimate is replaced by a more realistic one, the predicted parallax turns out to be too small to be detectable by Brahé's instruments.

A third example is a hypothetical one devised by Imre Lakatos (1970, pp. 100–1). It reads as follows:

> The story is about an imaginary case of planetary misbehaviour. A physicist of the pre-Einsteinian era takes Newton's mechanics and his law of gravitation, N, the accepted initial conditions, I, and calculates, with their help, the path of a newly discovered small planet, *p*. But the planet deviates from the calculated path.

Does our Newtonian physicist consider that the deviation was forbidden by Newton's theory and therefore that, once established, it refutes the theory N? No. He suggests that there must be a hitherto unknown planet p^1, which perturbs the path of p. He calculates the mass, orbit, etc. of this hypothetical planet and then asks an experimental astronomer to test his hypothesis. The planet p^1 is so small that even the biggest available telescopes cannot possibly observe it; the experimental astronomer applies for a research grant to build yet a bigger one. In three years time, the new telescope is ready. Were the unknown planet p^1 to be discovered, it would be hailed as a new victory of Newtonian science. But it is not. Does our scientist abandon Newton's theory and his idea of the perturbing planet? No. He suggests that a cloud of cosmic dust hides the planet from us. He calculates the location and properties of this cloud and asks for a research grant to send up a satellite to test his calculations. Were the satellite's instruments (possibly new ones, based on a little-tested theory) to record the existence of the conjectural cloud, the result would be hailed as an outstanding victory for Newtonian science. But the cloud is not found. Does our scientist abandon Newton's theory, together with the idea of the perturbing planet and the idea of the cloud which hides it? No. He suggests that there is some magnetic field in that region of the universe which disturbed the instruments of the satellite. A new satellite is sent up. Were the magnetic field to be found, Newtonians would celebrate a sensational victory. But it is not. Is this regarded as a refutation of Newtonian science? No. Either yet another ingenious auxiliary hypothesis is proposed or . . . the whole story is buried in the dusty volumes of periodicals and the story never mentioned again.

If this story is regarded as a plausible one, it illustrates how a theory can always be protected from falsification by deflecting the falsification to some other part of the complex web of assumptions.

Falsificationism inadequate on historical grounds

An embarrassing historical fact for falsificationists is that if their methodology had been strictly adhered to by scientists then those theories generally regarded as being among the best examples of scientific theories would never have been developed because they would have been rejected in their infancy. Given any example of a classic scientific theory, whether at the time of its first proposal or at a later date, it is possible to find observational claims that were generally accepted at the time and were considered to be inconsistent with the theory. Nevertheless, those theories were not rejected, and it is fortunate for science that they were not. Some historical examples to support my claim follow.

In the early years of its life, Newton's gravitational theory was falsified by observations of the moon's orbit. It took almost fifty years to deflect this falsification on to causes other than Newton's theory. Later in its life, the same theory was known to be inconsistent with the details of the orbit of the planet Mercury, although scientists did not abandon the theory for that reason. It turned out that it was never possible to explain away this falsification in a way that protected Newton's theory.

A second example concerns Bohr's theory of the atom, and is due to Lakatos (1970, pp. 140–54). Early versions of the theory were inconsistent with the observation that some matter is stable for a time that exceeds about 10^{-8} seconds. According to the theory, negatively charged electrons within atoms orbit around positively charged nuclei. But according to the classical electromagnetic theory presupposed by Bohr's theory, orbiting electrons should radiate. The radiation would result in an orbiting electron losing energy and collapsing into the nucleus. The quantitative details of classical electromagnetism yield an estimated time of about 10^{-8} seconds for this collapse to occur. Fortunately, Bohr persevered with his theory, in spite of this falsification.

A third example concerns the kinetic theory and has the advantage that the falsification of that theory at birth was explicitly acknowledged by its originator. When Maxwell (1965, vol. 1, p. 409) published the first details of a statistical version of the kinetic theory of gases in 1859, in that very same paper he acknowledged the fact that the theory was falsified by measurements on the specific heats of gases. Eighteen years later, commenting on the consequences of the kinetic theory, Maxwell (1877) wrote:

> Some of these, no doubt, are very satisfactory to us in our present state of opinion about the constitution of bodies, but there are others which are likely to startle us out of our complacency and perhaps ultimately to drive us out of all the hypotheses in which we have hitherto found refuge into that thoroughly conscious ignorance which is a prelude to every real advance in knowledge.

All the important developments within the kinetic theory took place after this falsification. Once again, it is fortunate that the theory was not abandoned in the face of falsifications by measurements of the specific heats of gases, as the naive falsificationist would be forced to insist.

A fourth example, the Copernican Revolution, will be outlined in more detail in the following section. This example emphasises the difficulties that arise for the falsificationist when the complexities of major theory changes are taken into account. The example also sets the scene for a discussion of some more recent and more adequate attempts to characterise the essence of science and its methods.

The Copernican Revolution

It was generally accepted in medieval Europe that the earth lies at the centre of a finite universe and that the sun, planets and stars orbit around it. The physics and cosmology that provided the framework in which this astronomy was set was basically that developed by Aristotle in the fourth century BC. In the second century AD, Ptolemy devised a detailed astronomical system that specified the orbits of the moon, the sun and all the planets.

In the early decades of the sixteenth century, Copernicus devised a new astronomy, an astronomy involving a moving earth, which challenged the Aristotelian and Ptolemaic systems. According to the Copernican view, the earth is not stationary at the centre of the universe but orbits the sun along with the planets. By the time Copernicus's idea had been substantiated, the Aristotelian world view had been replaced by the Newtonian one. The details of the story of this major theory change, a change that took place over one and a half centuries, do not lend support to the methodologies advocated by the inductivists and falsificationists, and indicate a need for a different, perhaps more complexly structured, account of science and its growth.

When Copernicus first published the details of his new astronomy, in 1543, there were many arguments that could be, and were, levelled against it. Relative to the scientific knowledge of the time, these arguments were sound ones and Copernicus could not satisfactorily defend his theory against them. In order to appreciate this situation, it is necessary to be familiar with some aspects of the Aristotelian world view on which the arguments against Copernicus were based. A very brief sketch of some of the relevant points follows.

The Aristotelian universe was divided into two distinct regions. The sub-lunar region was the inner region, extending from the central earth to just inside the moon's orbit. The super-lunar region was the remainder of the finite universe, extending from the moon's orbit to the sphere of the stars, which marked the outer boundary of the universe. Nothing existed beyond the outer sphere, not even space. Unfilled space is an impossibility in the Aristotelian system. All celestial objects in the super-lunar region were made of an incorruptible element called Ether. Ether possessed a natural propensity to move around the centre of the universe in perfect circles. This basic idea became modified and extended in Ptolemy's astronomy. Since observations of planetary positions at various times could not be reconciled with circular, earth-centred orbits, Ptolemy introduced further circles, called epicycles, into the system. Planets moved in circles, or epicycles, the centres of which moved in circles around the earth. The orbits could be further refined by adding epicycles to epicycles in such a way that the resulting system was compatible with observations of planetary positions and capable of predicting future planetary positions.

In contrast to the orderly, regular, incorruptible character of the super-lunar region, the sub-lunar region was marked by change, growth and decay, generation and corruption. All substances in the sub-lunar region were mixtures of four elements, air, earth, fire and water, and the relative proportions of elements in a mixture determined the properties of the substance so constituted.

Each element had a natural place in the universe. The natural place for earth was at the centre of the universe; for water, on the surface of the earth; for air, in the region immediately above the surface of the earth; and for fire, at the top of the atmosphere, close to the moon's orbit. Consequently, each earthly object would have a natural place in the sub-lunar region depending on the relative proportion of the four elements that it contained. Stones, being mostly earth, have a natural place near the centre of the earth, whereas flames, being mostly fire, have a natural place near to the moon's orbit, and so on. All objects have a propensity to move in straight lines, upwards or downwards, towards their natural place. Thus stones have a natural motion straight downwards, towards the centre of the earth, and flames have a natural motion straight upwards, away from the centre of the earth. All motions other than natural motions require a cause. For instance, arrows need to be propelled by a bow and chariots need to be drawn by horses.

These, then, are the bare bones of the Aristotelian mechanics and cosmology that were presupposed by contemporaries of Copernicus, and which were utilised in arguments against a moving earth. Let us look at some of the forceful arguments against the Copernican system.

Perhaps the argument that constituted the most serious threat to Copernicus was the so-called tower argument. It runs as follows. If the earth spins on its axis, as Copernicus had it, then any point on the earth's surface will move a considerable distance in a second. If a stone is dropped from the top of a tower erected on the moving earth, it will execute its natural motion and fall towards the centre of the earth. While it is doing so the tower will be sharing the motion of the earth, due to its spinning. Consequently, by the time the stone reaches the surface of the earth the tower will have moved around from the position it occupied at the beginning of the stone's downward journey. The stone should therefore strike the ground some distance from the foot of the tower. But this does not happen in practice. The stone strikes the ground at the base of the tower. It follows that the earth cannot be spinning and that Copernicus's theory is false.

Another mechanical argument against Copernicus concerns loose objects such as stones and philosophers resting on the surface of the earth. If the earth spins, why are such objects not flung from the earth's surface, as stones would be flung from the rim of a rotating wheel? And if the earth, as well as spinning, moves bodily around the sun, why doesn't it leave the moon behind?

Some arguments against Copernicus based on astronomical considerations have been mentioned earlier in this book. They involved the absence of parallax in the observed positions of the stars and the fact that Mars and Venus, as viewed by the naked eye, do not change size appreciably during the course of the year.

Because of the arguments I have mentioned, and others like them, the supporters of the Copernican theory were faced with serious difficulties. Copernicus himself was very much immersed in Aristotelian metaphysics and had no adequate response to them.

In view of the strength of the case against Copernicus, it might well be asked just what there was to be said in favour of the Copernican theory in 1543. The answer is, 'not very much'. The main attraction of the Copernican theory lay in the neat way it explained a number of features of planetary motion, which could be explained in the rival Ptolemaic theory only in an unattractive, artificial way. The features are the retrograde motion of the planets and the fact that, unlike the other planets, Mercury and Venus always remain in the proximity of the sun. A planet at regular intervals regresses, that is, stops its westward motion among the stars (as viewed from earth) and for a short time retraces its path eastward before continuing its journey westward once again. In the Ptolemaic system, retrograde motion was explained by the somewhat ad hoc manoeuvre of adding epicycles especially designed for the purpose. In the Copernican system, no such artificial move is necessary. Retrograde motion is a natural consequence of the fact that the earth and the planets together orbit the sun against the background of the fixed stars. Similar remarks apply to the problem of the constant proximity of the sun, Mercury and Venus. This is a natural consequence of the Copernican system once it is established that the orbits of Mercury and Venus are inside that of the earth. In the Ptolemaic system, the orbits of the sun, Mercury and Venus have to be artificially linked together to achieve the required result.

Thus there were some mathematical features of the Copernican theory that were in its favour. Apart from these, the two rival systems were more or less on a par as far as simplicity and accord with observations of planetary positions are concerned. Circular sun-centred orbits cannot be reconciled with observation, so that Copernicus, like Ptolemy, needed to add epicycles, and the total number of epicycles needed to produce orbits in accord with known observations was about the same for the two systems. In 1543 the arguments from mathematical simplicity that worked in favour of Copernicus could not be regarded as an adequate counter to the mechanical and astronomical arguments that worked against him. Nevertheless, a number of mathematically capable natural philosophers were to be attracted to the Copernican system, and their efforts to defend it became increasingly successful over the next hundred years or so.

The person who contributed most significantly to the defence of the Copernican system was Galileo. He did so in two ways. First, he used a telescope to observe the heavens, and in so doing he transformed the observational data that the Copernican theory was required to explain. Second, he devised the beginnings of a new mechanics that was to replace Aristotelian mechanics and with reference to which the mechanical arguments against Copernicus were defused.

When, in 1609, Galileo constructed his first telescopes and trained them on the heavens, he made dramatic discoveries. He saw that there were many stars invisible to the naked eye. He saw that Jupiter has moons and he saw that the surface of the earth's moon is covered with mountains and craters. He also observed that the apparent size of Mars and Venus, as viewed through the telescope, changed in the way predicted by the Copernican system. Later, Galileo was to confirm that Venus has phases like the moon, a fact that could be straightforwardly accommodated into the Copernican, but not the Ptolemaic, system. The moons of Jupiter defused the Aristotelian argument against Copernicus based on the fact that the moon stays with an allegedly moving earth. For now Aristotelians were faced with the same problem with respect to Jupiter and its moons. The earthlike surface of the moon undermined the Aristotelian distinction between the perfect, incorruptible heavens and the changing, corruptible earth.

The discovery of the phases of Venus marked a success for the Copernicans and a new problem for the Ptolemaics. It is undeniable that once the observations made by Galileo through his telescope are accepted, the difficulties facing the Copernican theory are diminished.

The foregoing remarks on Galileo and the telescope raise a serious epistemological problem. Why should observations through a telescope be preferred to naked-eye observations? One answer to this question might utilise an optical theory of the telescope that explains its magnifying properties and that also gives an account of the various aberrations to which we can expect telescopic images to be subject. But Galileo himself did not utilise an optical theory for that purpose. The first optical theory capable of giving support in this direction was devised by Galileo's contemporary, Kepler, early in the sixteenth century, and this theory was improved and augmented in later decades. A second way of facing our question concerning the superiority of telescopic to naked-eye observation is to demonstrate the effectiveness of the telescope in a practical way, by focusing it on distant towers, ships, etc. and demonstrating how the instrument magnifies and renders objects more distinctly visible. However, there is a difficulty with this kind of justification of the use of the telescope in astronomy. When terrestrial objects are viewed through a telescope, it is possible to separate the viewed object from aberrations contributed by the telescope because of the observer's familiarity with what a tower, a ship, etc. look like. This does not apply when an observer searches the heavens for he knows not what. It is significant in this respect that Galileo's drawing of the moon's surface as he saw it through a telescope contains some craters that do not in fact exist there. Presumably those 'craters' were aberrations arising from the functioning of Galileo's far-from-perfect telescopes. Enough has been said in this paragraph to indicate that the justification of telescopic observations was no simple, straightforward matter. Those adversaries of Galileo who queried his findings were not all stupid, stubborn reactionaries. Justifications were forthcoming, and became more and more adequate as better and better telescopes were constructed and as optical theories of their functioning were developed. But all this took time.

Galileo's greatest contribution to science was his work in mechanics. He laid some of the foundations of the Newtonian mechanics that was to replace Aristotle's. He distinguished clearly between velocity and acceleration and asserted that freely falling objects move with a constant acceleration that is independent of their weight, dropping a distance proportional to the square of the time of fall. He denied the Aristotelian claim that all motion requires a cause. He argued that the velocity of an object moving horizontally, along a line concentric with the earth, should neither increase nor decrease since it is neither rising nor falling. He analysed projectile motion by resolving the motion of a projectile into a horizontal component moving with a constant velocity and a vertical component subject to a constant acceleration downwards. He showed that the resulting path of a projectile was a parabola. He developed the concept of relative motion and argued that the uniform, horizontal motion of a system could not be detected by mechanical means without access to some reference point outside of the system.

These major developments were not achieved instantaneously by Galileo. They emerged gradually over a period of half a century, culminating in his book *Two New Sciences* (1974), which was first published in 1638, almost a century after the publication of Copernicus's major work. Galileo rendered his new conceptions meaningful and increasingly more precise by means of illustrations and thought experiments. Occasionally, Galileo described actual experiments, for instance experiments involving the rolling of spheres down inclined planes, although just how many of these Galileo actually performed is a matter of some dispute.

Galileo's new mechanics enabled the Copernican system to be defended against some of the objections to it mentioned above. An object held at the top of a tower and sharing with the tower a circular motion around the earth's centre will continue in that motion, along with the tower, after it is dropped and will consequently strike the ground at the foot of the tower, consistent with experience. Galileo took the argument further and claimed that the correctness of his views on horizontal motion could be demonstrated by dropping a stone from the top of the mast of a uniformly moving ship and noting that it strikes the deck at the foot of the mast, although Galileo did not claim to have performed the experiment. Galileo was less successful in explaining why loose objects are not flung from the surface of a spinning earth.

Although the bulk of Galileo's scientific work was designed to strengthen the Copernican theory, Galileo did not himself devise a detailed astronomy, and seemed to follow the Aristotelians in their preference for circular orbits. It was Galileo's contemporary, Kepler, who contributed a major breakthrough in that direction when he discovered that each planetary orbit could be represented by a single ellipse, with the sun at one focus. This eliminated the complex system of epicycles that both Copernicus and Ptolemy had found necessary. No similar simplification is possible in the Ptolemaic, earth-centred system. Kepler had at his disposal Tycho Brahé's recordings of planetary positions, which were more accurate than those available to Copernicus. After a painstaking analysis of the data, Kepler arrived at this three laws of planetary motion, that planets move in elliptical orbits around the sun, that a line joining a planet to the sun covers equal areas in equal times, and that the square of the period of a planet is proportional to the cube of its mean distance from the sun.

Galileo and Kepler certainly strengthened the case in favour of the Copernican theory. However, more developments were necessary before that theory was securely based on a comprehensive physics. Newton was able to take advantage of the work of Galileo, Kepler and others to construct the comprehensive physics that he published in his *Principia* in 1687. He spelt out a clear conception of force as the cause of acceleration rather than motion, a conception that had been present in a somewhat confused way in the writings of Galileo and Kepler. Newton replaced Galileo's views on inertia with his law of linear inertia, according to which bodies continue to move in straight lines at uniform speed unless acted on by a force. Another major contribution by Newton was of course his law of gravitation. This enabled Newton to explain the approximate correctness of Kepler's laws of planetary motion and Galileo's law of free fall. In the Newtonian system, the realms of the celestial bodies and of earthly bodies were unified, each set of bodies moving under the influence of forces according to

Newton's laws of motion. Once Newton's physics had been constituted, it was possible to apply it in detail to astronomy. It was possible, for instance, to investigate the details of the moon's orbit, taking into account its finite size, the spin of the earth, the wobble of the earth upon its axis, and so on. It was also possible to investigate the departure of the planets from Kepler's laws due to the finite mass of the sun, interplanetary forces, etc. Developments such as these were to occupy some of Newton's successors for the next couple of centuries.

The story I have sketched here should be sufficient to indicate that the Copernican Revolution did not take place at the drop of a hat or two from the Leaning Tower of Pisa. It is also clear that neither the inductivists nor the falsificationists give an account of science that is compatible with it. New concepts of force and inertia did not come about as a result of careful observation and experiment. Nor did they come about through the falsification of bold conjectures and the continual replacement of one bold conjecture by another. Early formulations of the new theory, involving imperfectly formulated novel conceptions, were persevered with and developed in spite of apparent falsifications. It was only after a new system of physics had been devised, a process that involved the intellectual and practical labour of many scientists over several centuries, that the new theory could be successfully matched with the results of observation and experiment in a detailed way. No account of science can be regarded as anywhere near adequate unless it can accommodate such factors.

Inadequacies of the falsificationist demarcation criterion and Popper's response

Popper made a seductive case for his criterion of demarcation between science and non- or pseudo-science. Scientific theories should be falsifiable, that is, they should have consequences that can be tested by observation or experiment. One weakness of this criterion, if unqualified, is that it is too easily satisfied and, in particular, satisfied by many knowledge claims that Popper, for one, would wish to classify as non-science. Astrologists do make claims that are falsifiable (and frequently falsified), while the horoscopes published in newspapers and journals do make falsifiable (as well as unfalsifiable) claims. The same 'Your Stars' newspaper column that yielded the (unfalsifiable) prediction that 'luck is possible in sporting speculation' quoted in chapter 5 also promised those whose birthday is on March 28 that 'a new lover will put a sparkle in your eye and improve social activities', a promise that is certainly falsifiable. Any fundamentalist brand of Christianity that insists that the Bible be taken literally is falsifiable. The claim in Genesis that God created the seas and populated them with fish would be falsified if there were no sea and/or no fish. Popper himself notes that Freudian theory, to the extent that it construes dreams as wish fulfilments, faces the threat of falsification by nightmares.

One response that the falsificationist can give to this observation is to note that theories must not only be falsifiable, but must also be not falsified. This might eliminate the claims of horoscopes to be scientific, and Popper argues that it eliminates Freudian theory. But this solution cannot be adopted too readily lest it eliminate everything that the falsificationists wish to retain as scientific, for we have seen that most scientific theories have their problems and clash with some accepted observation or other. So it becomes allowable, according to the sophisticated falsificationist, to modify theories in the face of apparent falsifications, and even to hang on to theories in spite of falsifications in the hope that the problem can be solved in the future. This kind of response is captured in the following passage from Popper (1974, p. 55), which is an attempt by him to confront difficulties of the kind I am raising here.

> I have always stressed the need for some dogmatism: the dogmatic scientist has an important role to play. If we give in to criticism too easily, we shall never find out where the real power of our theories lies.

It is my view that this passage is illustrative of the extent to which falsificationism faces severe difficulties in the light of the kinds of criticism raised in this chapter. The thrust of falsificationism is to emphasise the critical component of science. Our theories are to be subject to ruthless criticism so that the inadequate ones can be weeded out and replaced by more adequate ones. Faced with the problems surrounding the degree of definiteness with which theories can be falsified, Popper admits that it is often necessary to retain theories in spite of apparent falsifications. So although ruthless criticism is recommended, what would appear to be its opposite, dogmatism, has a positive role to play too. One might well wonder what is left of falsificationism once dogmatism is allowed a key role. Further, if both a critical and a dogmatic attitude can be condoned, then it is difficult to see what attitudes are ruled out. (It would be ironic if the highly qualified version of falsificationism became so weak as to rule out nothing, thereby clashing with the main intuition that led Popper to formulate it!)

Further reading

A range of criticisms of Popper's falsificationism are contained in Schilpp (1974). Criticism of all but the most sophisticated brand of falsificationism is marshalled in Lakatos (1970). Many of the points made in this chapter concerning the incompatibility of falsificationism with the Copernican Revolution were taken from Feyerabend (1975).

Lakatos and Musgrave (1970) contains articles that critically compare Popper's position with those of Thomas Kuhn, whose views are discussed in the next chapter. There are some finely tuned criticisms of Popper's position in Mayo (1996).

CHAPTER 8

Theories as structures I: Kuhn's paradigms

Theories as structures

The sketch of the Copernican Revolution outlined in the previous chapter suggests that the inductivist and falsificationist accounts of science are too piecemeal. Concentrating on the relationship between theories and individual observation statements or sets of them, they seem to fail to grasp the complexity of the mode of development of major theories. Since the 1960s it has become common to conclude from this that a more adequate account of science must proceed from an understanding of the theoretical frameworks in which scientific activity takes place. The next three chapters are concerned with three influential accounts of science that have resulted from an adoption of this approach. (In chapter 13 we will have reason to question whether the 'theory-dominated' view of science has gone too far.)

One reason why there is seen to be a need to view theories as structures stems from the history of science. Historical study reveals that the evolution and progress of major sciences exhibit a structure that is not captured by the inductivist and falsificationist accounts. The Copernican Revolution has already supplied us with an example. The notion can be further enhanced by reflecting on the fact that for a couple of centuries after Newton, physics was carried out in the Newtonian framework, until that framework was challenged by relativity and quantum theory at the beginning of the twentieth century. However, the historical argument is not the only reason why some have seen the need to concentrate on theoretical frameworks. A more general, philosophical argument is closely linked with the ways in which observation can be said to be theory-dependent. In chapter 1 it was stressed that observation statements must be expressed in the language of some theory. Consequently, it is argued, the statements, and the concepts figuring in them, will be as precise and informative as the theory in whose language they are formed is precise and informative. For instance, I think it will be agreed that the Newtonian concept of mass has a more precise meaning than the concept of democracy, say. It is plausible to suggest that the reason for the relatively precise meaning of the former stems from the fact that the concept plays a specific, well-defined role in a precise, closely knit theory, Newtonian mechanics. So, for instance, what counts as an adequate measurement of mass is dependent on the dictates of Newtonian theory. Measurement of mass via a spring balance is problematic because mass and weight are distinct, the latter being dependent on the distance of a massive object from the centre of gravity of the earth.

The dependence of the meaning of concepts on the structure of the theory in which they occur, and the dependence of the precision of the former on the precision and degree of coherence of the latter, can be made plausible by noting the limitations of some of the alternative ways in which a concept might be thought to acquire meaning. One such alternative is the view that concepts acquire their meaning by way of a definition. Definitions must be rejected as a fundamental way of establishing meanings because concepts can only be defined in terms of other concepts, the meanings of which are given. If the meanings of these latter concepts are themselves established by definition, it is clear that an infinite regress will result unless the meanings of some concepts are known by other means. A dictionary is useless unless we already know the meanings of many words. Newton could not define mass or force in terms of previously available concepts. It was necessary for him to transcend the limits of the old conceptual framework by developing a new one. A second alternative is the suggestion that concepts acquire their meaning by way of ostensive definition. We saw, in our discussion of a child learning the meaning of 'apple' in chapter 1, that this is difficult to sustain even in the case of an elementary notion like 'apple'. It is even more implausible when it comes to the definition of something like 'mass' in mechanics or 'electric field' in electromagnetism.

The claim that concepts derive their meaning at least in part from the role they play in a theory can be given support by the following historical reflections. Contrary to popular myth, experiment was by no means the key to Galileo's innovations in mechanics. Many of the 'experiments' he refers to in articulating his theory are thought experiments. This can appear paradoxical for those who see novel theories arising as a result of experiment, but it is quite comprehensible if it is accepted that precise experimentation can only be carried out if one has a precise theory capable of yielding predictions in the form of precise observation statements. Galileo, it might be argued, was in the process of making a major contribution to the building of a new mechanics that was to prove capable of supporting detailed experimentation at a later stage. It need not be surprising that his efforts involved thought experiments, analogies and illustrative metaphors rather than detailed experimentation. A case could be made to the effect that the typical history of a concept, whether it be 'chemical element', 'atom', 'the unconscious' or whatever, involves the initial emergence of the concept as a vague idea, followed by its gradual clarification as the theory in which it plays a part takes a more precise and coherent form. The emergence of the concept of an electric field can be construed in a way that supports such a view. When the concept was first introduced by Faraday in the first half of the nineteenth century it was very vague, and was articulated with the aid of mechanical analogies involving such things as stretched strings and metaphorical uses of such terms as 'tension', 'power' and 'force'. The field concept became increasingly better defined as the relationship between the electric field and other electromagnetic quantities became more clearly specified. Once Maxwell had introduced his displacement current, again with the aid of mechanical analogies, it was possible to bring great coherence to the theory in the form of Maxwell's equations, which clearly specified the interrelationship between all the electromagnetic quantities. It was not long before the ether, which had been considered to be the mechanical seat of the fields, could be dispensed with, leaving the fields as clearly defined concepts in their own right.

In this section I have attempted to construct a rationale for approaching science by way of the theoretical frameworks within which scientific work and argumentation take place. In this and the following two chapters we look at the work of three important philosophers of science who have pursued this idea.

Introducing Thomas Kuhn

Inductivist and falsificationist accounts of science were challenged in a major way by Thomas Kuhn (1970a) in his book *The Structure of Scientific Revolutions*, first published in 1962, and then republished with a clarificatory Postscript eight years later. His views have reverberated in the philosophy of science ever since. Kuhn started his academic career as a physicist and then turned his attention to the history of science. On doing so, he found that his preconceptions about the nature of science were shattered. He came to believe that traditional accounts of science, whether inductivist or falsificationist, do not bear comparison with historical evidence. Kuhn's account of science was subsequently developed as an attempt to give a theory more in keeping with the historical situation as he saw it. A key feature of his theory is the emphasis placed on the revolutionary character of scientific progress, where a revolution involves the abandonment of one theoretical structure and its replacement by another, incompatible one. Another important feature is the important role played by the sociological characteristics of scientific communities.

Kuhn's picture of the way a science progresses can be summarised by the following open-ended scheme:

pre-science – normal science – crisis – revolution – new normal science – new crisis –

The disorganised and diverse activity that precedes the formation of a science eventually becomes structured and directed when a single *paradigm* becomes adhered to by a scientific community. A paradigm is made up of the general theoretical assumptions and laws and the techniques for their application that the members of a particular scientific community adopt. Workers within a paradigm, whether it be Newtonian mechanics, wave optics, analytical chemistry or whatever, practise what Kuhn calls *normal science*. Normal scientists will articulate and develop the paradigm in their attempt to account for and accommodate the behaviour of some relevant aspects of the real world as revealed through the results of experimentation. In doing so, they will inevitably experience difficulties and encounter apparent falsifications. If difficulties of that kind get out of hand, a *crisis* state develops. A crisis is resolved when an entirely new paradigm emerges and attracts the allegiance of more and more scientists until eventually the original, problem-ridden paradigm is abandoned. The discontinuous change constitutes a *scientific revolution*. The new paradigm, full of promise and not beset by apparently insuperable difficulties, now guides new normal scientific activity until it too runs into serious trouble and a new crisis followed by a new revolution results.

With this resumé as a foretaste, let us look at the various components of Kuhn's scheme in more detail.

Paradigms and normal science

A mature science is governed by a single paradigm.[1] The paradigm sets the standards for legitimate work within the science it governs. It coordinates and directs the 'puzzle-solving' activity of the groups of normal scientists who work within it. The existence of a paradigm capable of supporting a normal science tradition is the characteristic that distinguishes science from non-science, according to Kuhn. Newtonian mechanics, wave optics and classical electromagnetism all constituted and perhaps constitute paradigms and qualify as sciences. Much of modern sociology lacks a paradigm and consequently fails to qualify as science.

As will be explained below, it is of the nature of a paradigm to belie precise definition. Nevertheless, it is possible to describe some of the typical components that go to make up a paradigm. Among the components will be explicitly stated fundamental laws and theoretical assumptions. Thus Newton's laws of motion form part of the Newtonian paradigm and Maxwell's equations form part of the paradigm that constitutes classical electromagnetic theory. Paradigms will also include standard ways of applying the fundamental laws to a variety of types of situation. For instance, the Newtonian paradigm will include methods of applying Newton's laws to planetary motion, pendulums, billiard-ball collisions, and so on. Instrumentation and instrumental techniques necessary for bringing the laws of the paradigm to bear on the real world will also be included in the paradigm. The application of the Newtonian paradigm in astronomy involves the use of a variety of approved kinds of telescope, together with techniques for their use and a variety of techniques for the correction of the data collected with their aid. A further component of paradigms consists of some very general, metaphysical principles that guide work within a paradigm. Throughout the nineteenth century the Newtonian paradigm was governed by an assumption something like, 'The whole of the physical world is to be explained as a mechanical system operating under the influence of various forces according to the dictates of Newton's laws of motion', and the Cartesian program in the seventeenth century involved the principle, 'There is no void and the physical universe is a big clockwork in which all forces take the form of a push'. Finally, all paradigms will contain some very general methodological prescriptions such as, 'Make serious attempts to match your paradigm with nature', or 'Treat failures in attempts to match a paradigm with nature as serious problems'.

Normal science involves detailed attempts to articulate a paradigm with the aim of improving the match between it and nature. A paradigm will always be sufficiently imprecise and open-ended to leave plenty of that kind of work to be done. Kuhn portrays normal science as a puzzle-solving activity governed by the rules of a paradigm. The puzzles will be of both a theoretical and an experimental nature. Within the Newtonian paradigm, for instance, typical theoretical puzzles involve devising mathematical techniques for dealing with the motion of a planet subject to more than one attractive force, and developing assumptions suitable for applying Newton's laws to the motion of fluids. Experimental puzzles included the improvement of the accuracy of telescopic observations and the development of experimental techniques capable of yielding reliable measurements of the gravitational constant. Normal scientists must presuppose that a paradigm provides the means for the solution of the puzzles posed within it. A failure to solve a puzzle is seen as a failure of the scientist rather than as an inadequacy of the paradigm. Puzzles that resist solution are seen as *anomalies* rather than as falsifications of a paradigm. Kuhn recognises that all paradigms will contain some anomalies (for example the Copernican theory and the apparent size of Venus or the Newtonian paradigm and the orbit of Mercury) and rejects all brands of falsificationism.

Normal scientists must be uncritical of the paradigm in which they work. It is only by being so that they are able to concentrate their efforts on the detailed articulation of the paradigm and to perform the esoteric work necessary to probe nature in depth. It is the lack of disagreement over fundamentals that distinguishes mature, normal science from the relatively disorganised activity of immature *pre-science*. According to Kuhn, the latter is characterised by total disagreement and constant debate over fundamentals, so much so that it is impossible to get down to detailed, esoteric work. There will be almost as many theories as there are workers in the field and each theoretician will be obliged to start afresh and justify his or her own particular approach. Kuhn offers optics before Newton as an example. There was a wide diversity of theories about the nature of light from the time of the ancients up to Newton. No general agreement was reached and no detailed, generally accepted theory emerged before Newton proposed and defended his particle theory. The rival theorists of the pre-science period disagreed not only over fundamental theoretical assumptions but also over the kinds of observational phenomena that were relevant to their theories. Insofar as Kuhn recognises the role played by a paradigm in guiding the search for and interpretation of observable phenomena, he accommodates the sense in which observation and experiment can be said to be theory-dependent.

Kuhn insists that there is more to a paradigm than what can be explicitly laid down in the form of explicit rules and directions. He invokes Wittgenstein's discussion of the notion of 'game' to illustrate some of what he means. Wittgenstein argued that it is not possible to spell out necessary and sufficient conditions for an activity to be a game. When one tries, one invariably finds an activity that one's definition includes but that one would not want to count as a game, or an activity that the definition excludes but that one would want to count as a game. Kuhn claims that the same situation exists with respect to paradigms. If one tries to give a precise and explicit characterisation of some paradigm in the history of science or in present-day science, it always turns out that some work within the paradigm violates the characterisation. However, Kuhn insists that this state of affairs does not render the concept of paradigm untenable any more than the similar situation with respect to 'game' rules out legitimate use of that concept. Even though there is no complete, explicit characterisation, individual scientists acquire knowledge of a paradigm through their scientific education. By solving standard problems, performing standard experiments and eventually by doing a piece of research under a supervisor who is already a skilled practitioner within the paradigm, an aspiring scientist becomes acquainted with the methods, the techniques and the standards of that paradigm. The aspiring scientist will be no more able to give an explicit account of the methods and skills he or she has acquired than a master-carpenter will be able to fully describe what lies behind his or her skills. Much of the normal scientist's knowledge will be *tacit*, in the sense developed by Michael Polanyi (1973).

Because of the way they are trained, and need to be trained if they are to work efficiently, typical normal scientists will be unaware of and unable to articulate the precise nature of the paradigm in which they work. However, it does not follow from this that a scientist will not be able to articulate the presuppositions involved in the paradigm should the need arise. Such a need will arise when a paradigm is threatened by a rival. In those circumstances, it will be necessary to attempt to spell out the general laws and metaphysical and methodological principles involved in a paradigm in order to defend them against the alternatives involved in the threatening new paradigm. The next section summarises Kuhn's account of how a paradigm can run into trouble and be replaced by a rival.

Crisis and revolution

Normal scientists work confidently within a well-defined area dictated by a paradigm. The paradigm presents them with a set of definite problems together with methods that they are confident will be adequate for the solution of the problems. If they blame the paradigm for any failure to solve a problem, they will be open to the same charges as the carpenter who blames his tools. Nevertheless, failures will be encountered and such failures can eventually attain a degree of seriousness that constitutes a serious crisis for the paradigm and may lead to the rejection of a paradigm and its replacement by an incompatible alternative.

The mere existence of unsolved puzzles within a paradigm does not constitute a crisis. Kuhn recognises that paradigms will always encounter difficulties. There will always be anomalies. It is only under special sets of conditions that the anomalies can develop in such a way as to undermine confidence in the paradigm. An anomaly will be regarded as particularly serious if it is seen as striking at the very fundamentals of a paradigm and yet persistently resists attempts by the members of the normal scientific community to remove it. Kuhn cites as an example problems associated with the ether and the earth's motion relative to it in Maxwell's electromagnetic theory, towards the end of the nineteenth century. A less-technical example would be the problems that comets posed for the ordered and full Aristotelian cosmos of interconnected crystalline spheres. Anomalies are also regarded as serious if they are important with respect to some pressing social need. The problems that beset Ptolemaic astronomy were pressing ones in the light of the need for calendar reform at the time of Copernicus. Also bearing on the seriousness of an anomaly will be the length of time that it resists attempts to remove it. The number of serious anomalies is a further factor influencing the onset of a crisis.

According to Kuhn, an analysis of the characteristics of a crisis period in science demands the competence of the psychologist as much as that of the historian. When anomalies come to be seen as posing serious problems for a paradigm, a period of 'pronounced professional insecurity' sets in. Attempts to solve the problem become more and more radical and the rules set by the paradigm for the solution of problems become progressively more loosened. Normal scientists begin to engage in philosophical and metaphysical disputes and try to defend their innovations, of dubious status from the point of view of the paradigm, by philosophical arguments. Scientists even begin to express openly their discontent with and unease over the reigning paradigm. Kuhn (1970a, p. 84) quotes Wolfgang Pauli's response to what he saw as the growing crisis in physics around 1924. An exasperated Pauli confessed to a friend, 'At the moment, physics is again terribly confused. In any case, it is too difficult for me, and I wish I had been a movie comedian or something of the sort and had never heard of physics'. Once a paradigm has been weakened and undermined to such an extent that its proponents lose their confidence in it, the time is ripe for revolution.

The seriousness of a crisis deepens when a rival paradigm makes its appearance. According to Kuhn (1970a, p. 91), 'the new paradigm, or a sufficient hint to permit later articulation, emerges all at once, sometimes in the middle of the night, in the mind of a man deeply immersed in crisis'. The new paradigm will be very different from and incompatible with the old one. The radical differences will be of a variety of kinds.

Each paradigm will regard the world as being made up of different kinds of things. The Aristotelian paradigm saw the universe as divided into two distinct realms, the incorruptible and unchanging super-lunar region and the corruptible and changing earthly region. Later paradigms saw the entire universe as being made up of the same kinds of material substances. Pre-Lavoisier chemistry involved the claim that the world contained a substance called phlogiston, which is driven from materials when they are burnt. Lavoisier's new paradigm implied that there is no such thing as phlogiston, whereas the gas, oxygen, does exist and plays a quite different role in combustion. Maxwell's electromagnetic theory involved an ether occupying all space, whereas Einstein's radical recasting of it eliminated the ether.

Rival paradigms will regard different kinds of questions as legitimate or meaningful. Questions about the weight of phlogiston were important for phlogiston theorists and vacuous for Lavoisier. Questions about the mass of planets were fundamental for Newtonians and heretical for Aristotelians. The problem of the velocity of the earth relative to the ether, which was deeply significant for pre-Einsteinian physicists, was dissolved by

Einstein. As well as posing different kinds of questions, paradigms will involve different and incompatible standards. Unexplained action at a distance was permitted by Newtonians but dismissed by Cartesians as metaphysical and even occult. Uncaused motion was nonsense for Aristotle and axiomatic for Newton. The transmutation of elements has an important place in modern nuclear physics (as it did in medieval alchemy and in seventeenth-century mechanical philosophy) but ran completely counter to the aims of Dalton's atomistic program. A number of kinds of events describable within modern microphysics involve an indeterminancy that had no place in the Newtonian program.

The way scientists view a particular aspect of the world will be guided by the paradigm in which they are working. Kuhn argues that there is a sense in which proponents of rival paradigms are 'living in different worlds'. He cites as evidence the fact that changes in the heavens were first noted, recorded and discussed by Western astronomers after the proposal of the Copernican theory. Before that, the Aristotelian paradigm had dictated that there could be no change in the super-lunar region and, accordingly, no change was observed. Those changes that were noticed were explained away as disturbances in the upper atmosphere.

The change of allegiance on the part of individual scientists from one paradigm to an incompatible alternative is likened by Kuhn to a 'gestalt switch' or a 'religious conversion'. There will be no purely logical argument that demonstrates the superiority of one paradigm over another and that thereby compels a rational scientist to make the change. One reason why no such demonstration is possible is the fact that a variety of factors are involved in a scientist's judgment of the merits of a scientific theory. An individual scientist's decision will depend on the priority he or she gives to the various factors. The factors will include such things as simplicity, the connection with some pressing social need, the ability to solve some specified kind of problem, and so on. Thus one scientist might be attracted to the Copernican theory because of the simplicity of certain mathematical features of it. Another might be attracted to it because in it there is the possibility of calendar reform. A third might have been deterred from adopting the Copernican theory because of an involvement with terrestrial mechanics and an awareness of the problems that the Copernican theory posed for it. A fourth might reject Copernicanism for religious reasons.

A second reason why no logically compelling demonstration of the superiority of one paradigm over another exists stems from the fact that proponents of rival paradigms will subscribe to different sets of standards and metaphysical principles. Judged by its own standards, paradigm *A* may be judged superior to paradigm *B*, whereas if the standards of paradigm *B* are used as premises, the judgment may be reversed. The conclusion of an argument is compelling only if its premises are accepted. Supporters of rival paradigms will not accept each others' premises and so will not necessarily be convinced by each others' arguments. It is for this kind of reason that Kuhn (1970a, pp. 93–4) compares scientific revolutions with political revolutions. Just as 'political revolutions aim to change political institutions in ways that those institutions themselves prohibit' and consequently 'political recourse fails', so the choice 'between competing paradigms proves to be a choice between incompatible modes of community life', and no argument can be 'logically or even probabilistically compelling'. This is not to say, however, that various arguments will not be among the important factors that influence the decisions of scientists. On Kuhn's view, the kinds of factors that do prove effective in causing scientists to change paradigms is a matter to be discovered by psychological and sociological investigation.

There are a number of interrelated reasons, then, why, when one paradigm competes with another, there is no logically compelling argument that dictates that a rational scientist should abandon one for the other. There is no single criterion by which a scientist must judge the merit or promise of a paradigm, and, further, proponents of competing programs will subscribe to different sets of standards and will even view the world in different ways and describe it in different languages. The aim of arguments and discussions between supporters of rival paradigms should be persuasion rather than compulsion. I suggest that what I have summarised in this paragraph is what lies behind Kuhn's claim that rival paradigms are 'incommensurable'.

A scientific revolution corresponds to the abandonment of one paradigm and the adoption of a new one, not by an individual scientist only but by the relevant scientific community as a whole. As more and more individual scientists, for a variety of reasons, are converted to the new paradigm, there is an 'increasing shift in the distribution of professional allegiances' (Kuhn, 1970a, p. 158). If the revolution is to be successful, this shift will spread so as to include the majority of the relevant scientific community, leaving only a few dissenters. These will be excluded from the new scientific community and will perhaps take refuge in a philosophy department. In any case, they will eventually die.

The function of normal science and revolutions

Some aspects of Kuhn's writings might give the impression that his account of the nature of science is a purely *descriptive* one, that is, that he aims to do nothing more than to describe scientific theories or paradigms and the activity of scientists. Were this the case, then Kuhn's account of science would be of little value as a *theory* of science. Unless the descriptive account of science is shaped by some theory, no guidance is offered as to what kinds of activities and products of activities are to be described. In particular, the activities and productions of hack scientists would need to be documented in as much detail as the achievements of an Einstein or a Galileo.

However, it is a mistake to regard Kuhn's characterisation of science as arising solely from a description of the work of scientists. Kuhn insists that his account constitutes a theory of science because it includes an explanation of the *function* of its various components. According to Kuhn, normal science and revolutions serve necessary functions, so that science must either involve those characteristics or some others that would serve to perform the same functions. Let us see what those functions are, according to Kuhn.

Periods of normal science provide the opportunity for scientists to develop the esoteric details of a theory. Working with a paradigm, the fundamentals of which they take for granted, they are able to perform the exacting experimental and theoretical work necessary to improve the match between the paradigm and nature to an ever-greater degree. It is through their confidence in the adequacy of a paradigm that scientists are able to devote their energies to attempts to solve the detailed puzzles presented to them within the paradigm, rather than engage in disputes about the legitimacy of their fundamental assumptions and methods. It is necessary for the normal scientist to be to a large extent uncritical. If all scientists were critical of all parts of the framework in which they worked all of the time then no detailed work would ever get done.

If all scientists were and remained normal scientists, a particular science would become trapped in a single paradigm and would never progress beyond it. This would be a serious fault, from the Kuhnian point of view. A paradigm embodies a particular conceptual framework through which the world is viewed and in which it is described, and a particular set of experimental and theoretical techniques for matching the paradigm with nature. But there is no *a priori* reason to expect that any one paradigm is perfect or even the best available. There are no inductive procedures for arriving at perfectly adequate paradigms. Consequently, science should contain within it a means of breaking out of one paradigm into a better one. This is the function of revolutions. All paradigms will be inadequate to some extent as far as their match with nature is concerned. When the mismatch becomes serious, that is, when a crisis develops, the revolutionary step of replacing the entire paradigm with another becomes essential for the effective progress of science.

Progress through revolutions is Kuhn's alternative to the cumulative progress characteristic of inductivist accounts of science. According to the latter view, scientific knowledge grows continuously as more numerous and more various observations are made, enabling new concepts to be formed, old ones to be refined, and new lawful relationships between them to be discovered. From Kuhn's particular point of view, this is mistaken, because it ignores the role played by paradigms in guiding observation and experiment. It is just because paradigms have such a pervasive influence on the science practised within them that the replacement of one by another must be a revolutionary one.

One other function catered for in Kuhn's account is worth

mentioning. Kuhn's paradigms are not so precise that they can be replaced by an explicit set of rules, as was mentioned above. Different scientists or groups of scientists may well interpret and apply the paradigm in a somewhat different way. Faced with the same situation, not all scientists will reach the same decision or adopt the same strategy. This has the advantage that the number of strategies attempted will be multiplied. Risks are thus distributed through the scientific community, and the chances of some long-term success are increased. 'How else', asks Kuhn (1970c, p. 241), 'could the group as a whole hedge its bets?'

The merits of Kuhn's account of science

There is surely something descriptively correct about Kuhn's idea that scientific work involves solving problems within a framework that is, in the main, unquestioned. A discipline in which fundamentals are constantly brought into question, as characterised in Popper's method of 'conjectures and refutations', is unlikely to make significant progress simply because principles do not remain unchallenged long enough for esoteric work to be done. It is all very well painting a heroic picture of Einstein as making a major advance by having the originality and courage to challenge some of the fundamental principles of physics, but we should not lose sight of the fact that it took two hundred years of detailed work within the Newtonian paradigm and one hundred years of work within theories of electricity and magnetism to reveal the problems that Einstein was to recognise and solve with his theories of relativity. It is philosophy, rather than science, that comes closest to being adequately characterised in terms of constant criticism of fundamentals.

If we compare the attempts by Kuhn and by Popper to capture the sense in which astrology differs from a science, it is Kuhn's account that is the more convincing, as Deborah Mayo (1996, chapter 2) has convincingly argued. From a Popperian perspective, astrology can be diagnosed as a non-science either because it is unfalsifiable, or because it is falsifiable and shown to be false. The first will not work because, as Kuhn (1970b) points out, even in the period during the Renaissance when astrology was practised seriously, astrologers did make predictions that were falsifiable, and indeed were frequently falsified. But this latter fact cannot be taken as sufficient to rule out astrology as a science lest physics, chemistry and biology are ruled out on similar grounds, for, as we have seen, all sciences have their problems in the form of problematic observations or experimental results. Kuhn's response is to suggest that the difference between, say, astronomy and astrology is that astronomers are in a position to learn from predictive failures in a way that astrologers are not. Astronomers can refine their instruments, test for possible disturbances, postulate undetected planets or lack of sphericity of the moon and so on and then carry out the detailed work to see if such changes can remove the problem posed by a failed prediction. Astrologers, by contrast, do not have the resources to learn from failures in the same way. But the 'resources' that astronomers have and astrologers lack can be interpreted as a shared paradigm that can sustain a normal science tradition. Kuhn's 'normal science', then, serves to identify a crucial element of a science.

The complementary part of Kuhn's account, 'scientific revolutions', would seem to be of considerable merit too. Kuhn used the notion of a revolution to stress the non-cumulative nature of the advance of science. The long-term progress of science does not merely involve the accumulation of confirmed facts and laws, but, on occasions also involves the overthrow of one paradigm and its replacement by an incompatible new one. Kuhn was certainly not the first to make this point. As we have seen, Popper himself stressed that scientific progress involves the critical overthrow of theories and their replacement by alternative ones. But, whereas for Popper the replacement of one theory by another is simply the replacement of one set of claims by a different set, there is much more to a scientific revolution from Kuhn's point of view. A revolution involves not merely a change in the general laws but also a change in the way the world is perceived and a change in the standards that are brought to bear in appraising a theory. As we have seen, the Aristotelian theory assumed a finite universe that was a system in which each item had a natural place and function, an important detail being the distinction between the celestial and the terrestrial. Within that scheme reference to the *function* of various items in the universe was a legitimate mode of explanation (for example, stones fall to the ground to reach their natural place and restore the universe to its ideal order). After the scientific revolution of the seventeenth century, the universe is an infinite one with items in it that interact by way of forces governed by laws. All explanations are by way of an appeal to those forces and laws. Insofar as empirical evidence played a role in the Aristotelian and Newtonian theories (or paradigms), in the former the evidence of the unaided senses operating under optimum conditions was regarded as fundamental, whereas in the latter, evidence acquired by way of instruments and experimentation was fundamental and often preferred over the direct deliverances of the senses.

Kuhn is undoubtedly correct, as a matter of descriptive fact, to note that there are such things as scientific revolutions that involve a change, not just in the range of claims made but also in the kind of entities that are assumed to make up the world and the kinds of evidence and modes of explanation that are deemed appropriate. What is more, once this is acknowledged, then any adequate account of scientific progress must include an account of how the changes made in the course of a revolution can be construed as progressive. Indeed, we can draw on Kuhn's characterisation of science and pose the problem in a particularly acute way. Kuhn insisted that what counts as a problem can change from paradigm to paradigm, and also that the standards of adequacy that are brought to bear on proposed solutions to problems also vary from paradigm to paradigm. But if it is the case that standards vary from paradigm to paradigm, then what standards can be appealed to in order to judge that a paradigm is better than, and so constitutes progress over, the paradigm it replaces? In precisely what sense can science be said to progress through revolutions?

Kuhn's ambivalence on progress through revolutions

Kuhn is notoriously ambiguous on the basic question we have posed and which his own work serves to highlight. After the publication of *The Structure of Scientific Revolutions* Kuhn was charged with having put forward a 'relativist' view of scientific progress. I take this to mean that Kuhn proposed an account of progress according to which the question of whether a paradigm is better or not than one that it challenges does not have a definitive, neutral answer, but depends on the values of the individual, group or culture that makes the judgment. Kuhn clearly was not comfortable with that charge and, in the Postscript that he added to the second edition of his book he attempted to distance himself from relativism. He wrote (1970a, p. 206), 'later scientific theories are better than earlier ones for solving puzzles in the often quite different environments to which they are applied. That is not a relativist's position, and it displays the sense in which I am a convinced believer in scientific progress'. This criterion is problematic insofar as Kuhn himself stresses that what counts as a puzzle and a solution to it is paradigm-dependent and also insofar as Kuhn (1970a, p. 154) elsewhere offers different criteria such as 'simplicity, scope and compatibility with other specialties'. But even more problematic is the clash between the non-relativist claim about progress and the numerous passages in Kuhn's book that read as an explicit advocacy of the relativist position, and even as a denial that there is a rational criterion of scientific progress at all.

Kuhn likens scientific revolutions to gestalt switches, to religious conversions and to political revolutions. Kuhn uses these comparisons to stress the extent to which the change of allegiance on the part of a scientist from one paradigm to another cannot be brought about by rational argument appealing to generally accepted criteria. The way in which the diagram on p. 6 changes from a staircase viewed from above to a staircase viewed from below is a modest example of a gestalt switch, but it serves to emphasise the extent to which such a switch is the very antithesis of a reasoned choice, and religious conversions are typically considered to be an analogous kind of change. As far as the analogy with political revolutions is concerned, Kuhn (1970a, pp. 93–4) insists that those revolutions 'aim to change political institutions in ways that those institutions themselves prohibit' so that 'political recourse fails'. By analogy, the choice 'between competing paradigms proves to be a choice between incompatible modes of community life' so that no argument can be 'logically or even probabilistically compelling'. Kuhn's insistence (1970a, p. 238) that the way in which we are to discover the nature of science is 'intrinsically sociological' and is to be accomplished by 'examining the nature of the scientific group, discovering what it values, what it tolerates and what it disdains', also leads to relativism if it transpires that different groups value, tolerate and disdain different things. This, indeed, is how proponents of the sociology of science currently in vogue commonly interpret Kuhn, developing his views into an explicit relativism.

In my view, Kuhn's account of scientific progress as it appears in the second edition of his book, complete with Postscript, contains two incompatible strands, one relativist and one not. This opens up two possibilities. The first is to follow the path taken by the sociologists mentioned in the previous paragraph and to embrace and develop the relativist strand in Kuhn's thought, which among other things involves carrying out the sociological investigation of science the need for which Kuhn alluded but never responded to. The second alternative is to ignore the relativism and rewrite Kuhn in a way that is compatible with some overarching sense of progress in science. This alternative will require an answer to the question of the sense in which a paradigm can be said to constitute progress over the one it replaces. I hope it will be clear by the end of the book which option I regard as the most fruitful.

Objective knowledge

'The transition between competing paradigms . . . must occur all at once (though not necessarily in an instant) or not at all.' I am not the only one to have found this sentence from Kuhn (1970a, p. 150) puzzling. How can a paradigm change take place all at once, but not necessarily in an instant? I do not think it is difficult to find the source of the confusion embodied in the problematic sentence. On the one hand, Kuhn is aware of the fact that a scientific revolution extends over a considerable period of time involving much theoretical and experimental work. Kuhn's own classic study of the Copernican Revolution (1959) documents the centuries of work involved. On the other hand, Kuhn's comparisons between paradigm change and gestalt switches or religious conversions make immediate sense of the idea that the change takes place 'all at once'. I suggest that Kuhn is, in effect, confusing two kinds of knowledge here, and it is important and helpful to spell out the distinction.

If I say 'I know the date on which I wrote this particular paragraph and you do not', I am referring to knowledge that I am acquainted with and that resides in my mind or brain, but which you are not acquainted with and is absent from your mind or brain. I know Newton's first law of motion but I do not know how to biologically classify a crayfish. Again, this is a question about what resides in my mind or brain. The claims that Maxwell was unaware that his electromagnetic theory predicted radio waves and that Einstein was aware of the results of the Michelson-Morley experiment involve this same usage of 'know' in the sense of 'being aware of'. Knowledge is a state of mind. Closely connected with this usage, in the sense that it is also to do with the states of mind of individuals, is the issue of whether or not, and the degree to which, an individual accepts or believes a claim or set of claims. I believe that Galileo made a convincing case for the validity of the use of his telescope, but Feyerabend did not. Ludwig Boltzmann accepted the kinetic theory of gases but his compatriot Ernst Mach did not. All these ways of talking about knowledge and claims to knowledge are about the states of mind or attitudes of individuals. It is a common and perfectly legitimate way of talking. For want of a better term I will call what is talked of here knowledge in the *subjective* sense. I will distinguish it from a different usage, which I refer to as knowledge in the *objective* sense.

The sentence 'my cat lives in a house that no animals inhabit' has the property of being contradictory, while the sentences 'I have a cat' and 'today a guinea pig died' have the property of being consequences of the statement 'today my white cat killed someone's pet guinea pig'. In these examples, the fact that the sentences have the properties I attribute to them, in some common sense, is obvious, but this need not be so. For example, a lawyer in a murder trial may, after much painstaking analysis, discover the fact that one witness's report has consequences that contradict those of a second witness. If that is indeed the case, then it is the case whether the witnesses in question were aware of it or believed it or not. What is more, if the lawyer had not discovered the inconsistency, it may have remained undiscovered, so that no one ever became aware of it. Nevertheless, it would remain the case that the statements were inconsistent. Propositions can have properties that are distinct from what individuals might be aware of. They have objective properties.

We have already encountered, in chapter 1, an instance of the distinction between subjective and objective knowledge. I drew a distinction between the perceptual experiences of individuals, and what they might believe as a consequence of them, on the one hand, and the observation statements that they might be taken to support on the other. I made the point that the latter are publicly testable and debatable in a way that the former are not.

The maze of propositions involved in a body of knowledge at some stage in its development will, in a similar way, have properties that individuals working on it need not be aware of. The theoretical structure that is modern physics is so complex that it clearly cannot be identified with the beliefs of any one physicist or group of physicists. Many scientists contribute in their separate ways and with their individual skills to the growth and articulation of physics, just as many workers combine their efforts in the construction of a cathedral. And just as a happy steeplejack may be blissfully unaware of the implication of some ominous discovery made by labourers digging near the foundations, so a lofty theoretician may be unaware of the relevance of some experimental finding for the theory on which he or she works. In either case, objective relationships exist between parts of the structure independently of whether individuals are aware of that relationship.

Historical examples from science that illustrate this point are easy to find. It is frequently the case that unexpected consequences of a theory, such as an experimental prediction or a clash with another theory, are *discovered* by subsequent work. Thus Poisson was able to discover and demonstrate that Fresnel's theory of light had the consequence that a bright spot should be visible at the centre of the shadow side of a suitably illuminated opaque disc, a consequence of which Fresnel had been unaware. Various clashes between Fresnel's theory and Newton's particle theory of light, which it challenged, were also discovered. For example, the former predicted that light should travel faster in air than in water, whereas the latter predicted the reverse.

I have illustrated a sense in which knowledge can be construed as objective by talking of the objective properties of statements, especially statements of theoretical and observational claims. But it is not only such statements that are objective. Experimental set-ups and procedures, methodological rules and mathematical systems are objective too, in the sense that they are distinct from the kinds of things that reside in individual minds. They can be confronted and can be exploited, modified and criticised by individuals. An individual scientist will be confronted by an objective situation – a set of theories, experimental results, instruments and techniques, modes of argument and the like – and it is these that the scientist must use in order to attempt to modify and improve the situation.

I do not intend my use of the term 'objective' to be evaluative. Theories that are inconsistent or which explain little will be objective according to my usage. Indeed, such theories will objectively possess the properties of being inconsistent or explaining little. Although my usage of 'objective' derives from and follows closely that of Karl Popper (see especially his 1979 text, chapters 3 and 4), I do not wish to follow him in getting involved in the tricky question of the precise sense in which these objective properties exist. Statements do not have properties in the sense that physical objects do, and spelling out the mode of existence of such linguistic objects, as well as other social constructions such as methodological rules and mathematical systems, is a tricky philosophical business. I am content to make my points at a commonsense level, using the kinds of examples I have used. This is sufficient for my purpose.

Much of Kuhn's talk of paradigms fits well into the objective side of dichotomy I have introduced. His talk of the puzzle-solving tradition within a paradigm and the anomalies confronted by a paradigm, and also the way in which paradigms differ in involving different standards and different metaphysical assumptions, are all cases in point. Accepting this mode of talk, it is quite meaningful, in Kuhn's terms, to formulate our basic question concerning the sense in which a particular paradigm can be said to be an improvement on its rival. This is a question about the objective relation between paradigms.

However, there is this other mode of talking at work in Kuhn's book which is situated on the subjective side of my dichotomy. This includes his talk of gestalt switches and the like. Talking of the switch from one paradigm to another in terms of gestalt switches, as Kuhn does, creates the impression that the viewpoints on either side of the switch cannot be compared. The change from one paradigm to another is identified with the change that takes place within a scientist's mind or brain when he or she changes allegiance from one to the other. It is this identification that leads to the confusion embodied in the sentence from Kuhn introduced at the beginning of this section. If our concern is the nature of science and the sense in which science can be said to progress, as Kuhn's seems to be, then my suggestion is that all the talk of gestalt switches and religious conversions be removed from Kuhn's account and that we stick to an objective characterisation of paradigms and the relationship between them. Much of the time Kuhn does precisely this, and his historical studies are a mine of important material for helping to elucidate the nature of science.

The way in which one historically existing paradigm might be said to be better than the rival that it replaces is distinct from the question of the ways in which, or the reasons why, individual scientists change their allegiance from one to the other, or come to be working in one or the other. The fact that individual scientists in their scientific work make judgments and choices for a variety of reasons, often influenced by subjective factors, is one thing. The relationship between one paradigm and another, perceivable most clearly with the benefit of hindsight, is another. If some distinctive sense in which science progresses is to be identified, it is the latter kind of consideration that will yield the answer. That is why I am dissatisfied with Kuhn's attempt, in his 1977 text (chapter 13), to combat the charge of relativism by focusing on 'value judgment and theory choice'.

Further reading

The key source is, of course, Kuhn's *The Structure of Scientific Revolutions* (1970a). In 'Logic of Discovery or Psychology of Research' (1970b) Kuhn discusses the relationship between his views and

Popper's and replies to some of his critics in 'Reflections on My Critics' (1970c). A valuable collection of Kuhn's essays is his 1977 text. A detailed discussion of Kuhn's philosophy of science is Hoyningen-Huene (1993), which contains a detailed bibliography of Kuhn's work. Bird (2000) is a more recent introduction to Kuhn's philosophy of science. Lakatos and Musgrave (1970) contains a number of interchanges between Kuhn and his critics. For appropriations of Kuhn's ideas by sociologists see, for example, Bloor (1971) and Barnes (1982). For an account of the construction of meaning in science that exemplifies the position outlined in the first section of this chapter, see Nersessian (1984).

CHAPTER 9

Theories as structures II: research programs

Introducing Imre Lakatos

Imre Lakatos was a Hungarian who moved to England in the late 1950s and came under the influence of Karl Popper who, in Lakatos's own words 'changed [his] life' (Worrall and Currie, 1978a, p. 139). Although an avid supporter of Popper's approach to science, Lakatos came to realise some of the difficulties that faced Popper's falsificationism, difficulties of the kind we have considered in chapter 7. By the mid-1960s Lakatos was aware of the alternative view of science contained in Kuhn's *The Structure of Scientific Revolutions.* Although Popper and Kuhn proposed rival accounts of science, their views do have much in common. In particular, they both take a stand against positivist, inductivist accounts of science. They both give priority to theory (or paradigm) over observation, and insist that the search for, interpretation and acceptance or rejection of the results of observation and experiment take place against a background of theory or paradigm. Lakatos carried on that tradition, and looked for a way of modifying Popper's falsificationism and ridding it of its difficulties, among other ways by drawing on some of the insights of Kuhn while totally rejecting the relativist aspects of the latter's position. Like Kuhn, Lakatos saw the merit in portraying scientific activity as taking place in a framework, and coined the phrase 'research program' to name what were, in a sense, Lakatos's alternatives to Kuhn's paradigms. The primary source for an account of Lakatos's methodology is his 1970 text.

Lakatos's research programs

We saw in chapter 7 that one of the main difficulties with Popper's falsificationism was that there was no clear guidance concerning which part of a theoretical maze was to be blamed for an apparent falsification. If it is left to the whim of the individual scientist to place the blame wherever he or she might wish, then it is difficult to see how the mature sciences could progress in the coordinated and cohesive way that they seem to do. Lakatos's response was to suggest that not all parts of a science are on a par. Some laws or principles are more basic than others. Indeed, some are so fundamental as to come close to being the defining feature of a science. As such, they are not to be blamed for any apparent failure. Rather, the blame is to be placed on the less fundamental components. A science can then be seen as the programmatic development of the implications of the fundamental principles. Scientists can seek to solve problems by modifying the more peripheral assumptions as they see fit. Insofar as their efforts are successful they will be contributing to the development of the same *research program* however different their attempts to tinker with the peripheral assumptions might be.

Lakatos referred to the fundamental principles as the *hard core* of a research program. The hard core is, more than anything else, the defining characteristic of a program. It takes the form of some very general hypotheses that form the basis from which the program is to develop. Here are some examples. The hard core of the Copernican program in astronomy was the assumption that the earth and the planets orbit a stationary sun and that the earth spins on its axis once a day. The hard core of Newtonian physics is comprised of Newton's three laws of motion plus his law of gravitational attraction. The hard core of Marx's historical materialism would be something like the assumption that major social change is to be explained in terms of class struggle, the nature of the classes and the details of the struggle being determined, in the last instance, by the economic base.

The fundamentals of a program need to be augmented by a range of supplementary assumptions in order to flesh it out to the point where definite predictions can be made. It will consist not only of explicit assumptions and laws supplementing the hard core, but also assumptions underlying the initial conditions used to specify particular situations and theories presupposed in the statement of observations and experimental results. For example, the hard core of the Copernican program needed to be supplemented by adding numerous epicycles to the initially circular orbits and it was also necessary to alter previous estimates of the distance of the stars from earth. Initially the program also involved the assumption that the naked eye serves to reveal accurate information concerning the position, size and brightness of stars and planets. Any inadequacy in the match between an articulated program and observation is to be attributed to the supplementary assumptions rather than the hard core. Lakatos referred to the sum of the additional hypotheses supplementing the hard core as the *protective belt*, to emphasise its role of protecting the hard core from falsification. According to Lakatos (1970, p. 133), the hard core is rendered unfalsifiable by 'the methodological decisions of its protagonists'. By contrast, assumptions in the protective belt are to be modified in an attempt to improve the match between the predictions of the program and the results of observation and experiment. For instance, the protective belt within the Copernican program was modified by substituting elliptical orbits for Copernicus's sets of epicycles and telescopic data for naked-eye data. The initial conditions also came to be modified eventually, with changes in the estimate of the distance of the stars from the earth and the addition of new planets. Lakatos made free use of the term 'heuristic' in characterising research programs. A heuristic is a set of rules or hints to aid discovery or invention. For example, part of a heuristic for solving crossword puzzles might be 'start with the clues requiring short-word answers and then proceed to those requiring long-word answers'. Lakatos divided guidelines for work within research programs into a *negative heuristic* and a *positive heuristic.* The negative heuristic specifies what the scientist is advised not to do. As we have already seen, scientists are advised not to tinker with the hard core of the program in which they work. If a scientist does modify the hard core then he or she has, in effect, opted out of the program. Tycho Brahé opted out of the Copernican program when he suggested that only the planets, but not the earth, orbit the sun and that the sun orbits the earth.

The *positive heuristic* of a program, that which specifies what scientists should do rather than what they should not do within a program, is more difficult to characterise specifically than the negative heuristic. The positive heuristic gives guidance on how the hard core is to be supplemented and how the resulting protective belt is to be modified in order for a program to yield explanations and predictions of observable phenomena. In Lakatos's own words (1970, p. 135), 'the positive heuristic consists of a partially articulated set of suggestions or hints on how to change, develop, the "refutable variants" of the research program, how to modify, sophisticate, the "refutable" protective belt'. The development of the program will involve not only the addition of suitable auxiliary hypotheses but also the development of adequate experimental and mathematical techniques. For instance, from the very inception of the Copernican program it was clear that mathematical techniques for combining and manipulating epicycles and improved techniques for observing planetary positions were necessary. Lakatos illustrated the notion of a positive heuristic with the story of Newton's early development of his gravitational theory. Here, the positive heuristic involved the idea that one should start with simple, idealised cases and then, having mastered them, one should proceed to more complicated, and more realistic, cases. Newton first arrived at the inverse square law of attraction by considering the elliptical motion of a point planet around a stationary point sun. It was clear that if the program was to be applied in practice to planetary motions then it would need to be developed from this idealised form to a more realistic one. But that development involved the solution of theoretical problems and was not to be achieved without considerable theoretical labour. Newton himself, faced with a definite program, that is, guided by his positive heuristic, made considerable progress. He first took into account the fact that the sun as well as a planet moves under the influence of their mutual attraction. Then he took account of the finite size of the planets and treated them as spheres. After solving the mathematical problem posed by that move, Newton proceeded to allow for other complications such as those introduced by the possibility that a planet can spin, and the fact that there are gravitational forces between the individual planets as well as between each planet and the sun. Once Newton had progressed that far in the program, following a path that had presented itself as more or less necessary from the outset, he began to be concerned about the match between his theory and observation. When the match was found wanting he was able to proceed to non-spherical planets and so on. As well as the theoretical program, the positive heuristic contained an experimental one. That program included the development of more accurate telescopes, together with auxiliary theories necessary for their use in astronomy, such as those providing adequate means for allowing for refraction of light in the earth's atmosphere. The initial formulation of Newton's program already indicated the desirability of constructing apparatus sensitive enough to detect gravitational attraction on a laboratory scale (Cavendish's experiment).

The program that had Newton's laws of motion and his law of gravitation at its core gave strong heuristic guidance. That is, a fairly definite program was mapped out from the start. Lakatos (1970, pp. 140–55) gives an account of the development of Bohr's theory of the atom as another example of a positive heuristic in action. An important feature of these examples of developing research programs, stressed by Lakatos, is the comparatively late stage at which observational testing becomes relevant. This is in keeping with the comments about Galileo's construction of his mechanics in the first section of chapter 8. Early work in a research program is portrayed as taking place without heed or in spite of apparent falsifications by observation. A research program must be given a chance to realise its full potential. A suitable sophisticated and adequate protective belt must be constructed. In our example of the Copernican program, this included the development of an adequate mechanics that could accommodate the earth's motion and an adequate optics to help interpret the telescopic data. When a program has been developed to the stage where it is appropriate to subject it to experimental tests, it is confirmations rather than falsifications that are of paramount significance, according to Lakatos. The worth of a research program is indicated by the extent to which it leads to novel predictions that are confirmed. The Newtonian program experienced dramatic confirmations of this kind when Galle first observed the planet Neptune and when

Halley's comet returned as predicted. Failed predictions, such as Newton's early calculations of the moon's orbit, are simply indications that more work needs to be done on supplementing or modifying the protective belt.

The main indication of the merit of a research program is the extent to which it leads to novel predictions that are confirmed. A second indication, implicit in our discussion above, is that a research program should indeed offer a *program* of research. The positive heuristic should be sufficiently coherent to be able to guide future research by mapping out a program. Lakatos suggested Marxism and Freudian psychology as programs that lived up to the second indicator of merit but not to the first, and contemporary sociology as one that lives up to the first to some extent but not the second (although he did not back up these remarks with any detail). In any event, a *progressive* research program will be one that retains its coherence and at least intermittently leads to novel predictions that are confirmed, while a *degenerating* program will be one that loses its coherence and/or fails to lead to confirmed novel predictions. The replacement of a degenerating program by a progressive one constitutes Lakatos's version of a scientific revolution.

Methodology within a program and the comparison of programs

We need to discuss Lakatos's methodology of scientific research programs in the context of work within a program and in the context of the clash between one research program and another. Work within a single research program involves the expansion and modification of its protective belt by the addition and articulation of various hypotheses. Any such move is permissible so long as it is not ad hoc in the sense discussed in chapter 6. Modifications or additions to the protective belt of a research program must be independently testable. Individual scientists or groups of scientists are open to modify or augment the protective belt in any way they choose, provided these moves open up the opportunity for new tests and hence the possibility of novel discoveries.

By way of illustration, let us take an example from the

development of the Newtonian program that we have employed several times before and consider the situation that confronted Leverrier and Adams when they addressed themselves to the troublesome orbit of the planet Uranus. Those scientists chose to modify the protective belt of the program by proposing that the initial conditions were inadequate and suggesting that there was an as yet unidentified planet close to Uranus and disturbing its orbit. Their move was in accordance with Lakatos's methodology because it was testable. The conjectured planet could be sought for by training telescopes on the appropriate region of the sky. But other possible responses would be legitimate according to Lakatos's position. For instance, the problematic orbit could be blamed on some new type of aberration of the telescope, provided the suggestion was made in a way that made it possible to test for the reality of such aberrations. In a sense, the more testable moves that are made to solve a problem such as this the better, because this increases the chances of success (where success means the confirmation of the novel predictions ensuing from a move). Moves that are ad hoc are ruled out by Lakatos's methodology. So, in our example, an attempt to accommodate Uranus's problematic orbit by simply labelling that complex orbit as the natural motion of Uranus would be ruled out. It opens up no new tests and hence no prospect of novel discoveries.

A second kind of move ruled out by Lakatos's methodology is one that involves a departure from the hard core. Making such a move destroys the coherence of a program and amounts to opting out of that program. For instance, a scientist attempting to cope with Uranus's orbit by suggesting that the attraction between Uranus and the sun was something other than the inverse square law would be opting out of the Newtonian research program.

The fact that any part of a complex theoretical maze might be responsible for an apparent falsification poses a serious problem for the falsificationist relying on an unqualified method of conjectures and refutations. For that person, the inability to locate the source of the trouble leads to unmethodical chaos. Lakatos's methodology is designed to avoid that consequence. Order is maintained by the inviolability of the hard core of the program and by the positive heuristic that accompanies it. The proliferation of ingenious conjectures within that framework will lead to progress provided some of the predictions resulting from those conjectures occasionally prove successful. Decisions to retain or reject a hypothesis are fairly straightforwardly determined by the results of experimental tests. The bearing of observation on a hypothesis under test is relatively unproblematic within a research program because the hard core and the positive heuristic serve to define a fairly stable background.

As was mentioned above, Lakatos's version of a Kuhnian revolution involves the ousting of one research program by another. We have seen that Kuhn (1970, p. 94) was unable to give a clear answer to the question of the sense in which a paradigm can be superior to the one it replaces, and so left him with no option but to appeal to the authority of the scientific community. Later paradigms are superior to their predecessors because the scientific community judges them to be so, and 'there is no standard higher than the assent of the relevant community'. Lakatos was dissatisfied with the relativist implications of Kuhn's theory. He sought a standard that lay outside of particular paradigms or, in Lakatos's case, research programs, which could be used to identify some non-relativist sense in which science progresses. To the extent that he had such a standard, it lay in his conception of progressing and degenerating research programs. Progress involves the replacement of a degenerating program with a progressive one, with the latter being an improvement on the former in the sense that it has been shown to be a more efficient predictor of novel phenomena.

Novel predictions

The non-relativist measure of progress that Lakatos proposed relied heavily on the notion of a novel prediction. One program is superior to another insofar as it is a more successful predictor of novel phenomena. As Lakatos came to realise, the notion of a novel prediction is not as straightforward as it might at first appear, and care is needed to mould that notion into a form that serves the purpose required of it within Lakatos's methodology or, indeed, any methodology that seeks to make significant use of it.

We have already met novel predictions in the context of Popper's methodology. In that context I suggested that the essence of

Popper's position is that a prediction is novel, at a particular time, to the extent that it does not figure in, or perhaps clashes with, the knowledge that is familiar and generally accepted at that time. For Popper, testing a theory by way of its novel predictions amounted to a severe test of that theory just because the prediction clashed with prevailing expectations. Lakatos's use of novel predictions in something like the Popperian sense to help him characterise the progressiveness of a research program will not do, as he himself came to realise, and this can be established by means of fairly straightforward counter examples, examples drawn from the very programs that Lakatos freely utilised to illustrate his position. The counter examples involve situations where the worth of a research program is demonstrated by its ability to explain phenomena that at the time were already well established and familiar, and so not novel in the Popperian sense.

There are features of planetary motion that have been well known since antiquity, but which were adequately explained only with the advent of the Copernican theory. They include the retrograde motion of the planets and the fact that the planets appear brightest when they are retrogressing, as well as the fact that Venus and Mercury never appear far from the sun. The qualitative features of these phenomena follow straightforwardly once it is assumed that the earth orbits the sun along with the planets and that the orbits of Mercury and Venus are inside that of the earth, whereas in the Ptolemaic theory they can only be explained by introducing epicycles designed specifically for the purpose. Lakatos joined Copernicus, and I imagine most of the rest of us, in recognising this as a major mark of the superiority of the Copernican over the Ptolemaic system. However, the Copernican prediction of the general features of planetary motion did not count as novel in the sense we have defined it for the straightforward reason that those phenomena had been well known since antiquity. The observation of parallax in the stars was probably the first confirmation of the Copernican theory by a prediction that counts as novel in the sense we are discussing, but that doesn't suit Lakatos's purpose at all, since it did not occur until well into the nineteenth century, well after the superiority of Copernicus over Ptolemy had been accepted within science.

Other examples are readily found. One of the few observations that could be invoked to support Einstein's general theory of relativity was the precession of the perihelion of the orbit of the planet Mercury, a phenomenon well known and accepted long before Einstein's theory explained it. One of the most impressive features of quantum mechanics was its ability to explain the spectra exhibited by the light emitted from gases, a phenomenon familiar to experimenters for over half a century before the quantum mechanical explanation was available. These successes can be described as involving the novel prediction of phenomena rather than the prediction of novel phenomena.

Lakatos came to realise, in the light of some considerations put forward by E. Zahar (1973), that the account of novel predictions in his original formulation of the methodology of scientific research programs needed to be modified. After all, when assessing the extent to which some observable phenomena support a theory or program, surely it is a historically contingent fact of no philosophical relevance whether it is the theory or knowledge of the phenomena that comes first. Einstein's theory of relativity can explain the orbit of Mercury and also the bending of light rays in a gravitational field. These are both considerable achievements that support the theory. It so happens that the precession of the perihelion of Mercury was known prior to Einstein's formulation of the theory, whereas the bending of light rays was discovered subsequently. But would it make any difference to our assessment of Einstein's theory if it had been the other way around, or if both phenomena had been known before or both discovered after? The fine details of the appropriate response to these reflections are still being debated, for example by Alan Musgrave (1974b) and John Worrall (1985 and 1989a), but the intuition that needs to be grasped, and which is at work in the comparison of Copernicus and Ptolemy, seems straightforward enough. The Ptolemaic explanation of retrograde motion did not constitute significant support for that program because it was artificially fixed up to fit the observable data by adding epicycles especially designed for the purpose. By contrast, the observable phenomena followed in a natural way from the fundamentals of the Copernican theory without any artificial adjustment. The predictions of a theory or program that count are those that are natural rather than contrived. Perhaps what lies behind the intuition here is the idea that evidence supports a theory if, without the theory, there are unexplained coincidences contained in the evidence. How could the Copernican theory successfully predict all the observable general features of planetary motion if it wasn't essentially correct? The same argument does not work in the case of the Ptolemaic explanation of the same phenomena. Even if the Ptolemaic theory is quite wrong, it is no coincidence that it can explain the phenomena because the epicycles have been added in such a way as to ensure that it does. This is the way in which Worrall (1985, 1989) treats the matter.

In the light of this, we should reformulate Lakatos's methodology so that a program is progressive to the extent that it makes natural, as opposed to novel, predictions that are confirmed, where 'natural' stands opposed to 'contrived' or 'ad hoc'. (We shall revisit this issue from a different and perhaps superior angle in chapter 13.)

Testing the methodology against history

Lakatos shared Kuhn's concern with the history of science. He believed it to be desirable that any theory of science be able to make sense of the history of science. That is, there is a sense in which a methodology or philosophy of science is to be tested against the history of science. However, the precise way in which this is so needs to be carefully spelt out, as Lakatos was well aware. If the need for a philosophy of science to match the history of science is interpreted undiscriminatingly, then a good philosophy of science will become nothing more than an accurate description of science. As such, it will be in no position to capture the essential characteristics of science or to discriminate between good science and bad science. Popper and Lakatos tended to regard Kuhn's account as 'merely' descriptive, in this sense, and hence deficient. Popper was so wary of the problem that he, unlike Lakatos, denied that comparison with the history of science was a legitimate way of arguing for a philosophy of science.

I suggest that the essentials of Lakatos's position, as described in his 1978 text, are these. There are episodes in the history of science that are unproblematically progressive and which can be recognised as such prior to any sophisticated philosophy of science. If someone wants to deny that Galileo's physics was an advance on Artistotle's or that Einstein's was an advance on Newton's then he or she is just not using the word science in the way that the rest of us are. To be concerned with the question of how best to categorise science we must have some pre-theoretical notion of what science is in order to formulate the question, and that pre-theoretical notion will include the ability to recognise classic examples of major scientific achievements such as those of Galileo and Einstein. With these presuppositions as a background, we can now demand that any philosophy or methodology of science be compatible with them. That is, any philosophy of science should be able to grasp the sense in which Galileo's achievements in astronomy and physics were in the main major advances. So if the history of science reveals that in his astronomy Galileo transformed what were considered to be the observable facts, and in his mechanics he relied mainly on thought experiments rather than real ones, then that poses a problem for those philosophies that portray scientific progress as cumulative, progressing by way of the accumulation of secure observational facts and cautious generalisations from them. Lakatos's own early version of his methodology of research programs can be criticised for utilising a notion of novel prediction in a way that makes it impossible to grasp the sense in which Copernicus's astronomy was progressive, as I did in the previous section.

With this mode of argument, Lakatos proceeds to criticise positivist and falsificationist methodologies on the grounds that they fail to make sense of classic episodes in the progress of science, and argues, by contrast, that his own account does not suffer from the same deficiency. Turning, then, to more minor episodes in the history of science, Lakatos, or a supporter, can pick on episodes from the history of science that have puzzled historians and philosophers and show how they make complete sense from the point of view of the methodology of scientific research programs. Thus, for example, many have been puzzled by the fact that when Thomas Young proposed the wave theory of light in the early nineteenth century it won few supporters, whereas Fresnel's version, devised two decades later, won widespread acceptance. John Worrall (1976) gives historical support to Lakatos's position when he shows that, as a matter of historical fact, Young's theory was not strongly confirmed experimentally in a natural, as opposed to a contrived, way, as Fresnel's was, and that Fresnel's version of the wave theory had a vastly superior positive heuristic by virtue of the mathematical tools he was able to introduce. A number of Lakatos's students or former students carried out studies, appearing in Howson (1976), intended to support Lakatos's methodology in this kind of way.

Lakatos came to see the main virtue of his methodology to be the aid it gives to the writing of the history of science. The historian must attempt to identify research programs, characterise their hard cores and protective belts, and document the ways in which they progressed or degenerated. In this way, light can be shed on the way science progresses by way of the competition between programs. I think it must be conceded that Lakatos and his followers did succeed in casting useful light on some classic episodes in the history of the physical sciences by studies carried out in this way, as the essays in Howson (1976) reveal. Although Lakatos's methodology can offer advice to historians of science, it was not intended by Lakatos as a source of advice for scientists. This became an inevitable conclusion for Lakatos given the way he found it necessary to modify falsificationism to overcome the problems it faced. Theories should not be rejected in the face of apparent falsifications because the blame might in due course be directed at a source other than the theory, and single successes certainly do not establish the merit of a theory for all time. That is why Lakatos introduced research programs, which are given time to develop and may come to progress after a degenerating period, or degenerate after early successes. (It is worth recalling in this connection that the Copernican theory degenerated for about a century after its early successes before the likes of Galileo and Kepler brought it to life again.) But once this move is taken, it is clear that there can be no on-the-spot advice forthcoming from Lakatos's methodology along the lines that scientists must give up a research program, or prefer a particular research program to its rival. It is not irrational or necessarily misguided for a scientist to remain working on a degenerating program if he or she thinks there are possible ways to bring it to life again. It is only in the long term (that is, from a historical perspective) that Lakatos's methodology can be used to meaningfully compare research programs. In this connection, Lakatos came to make a distinction between the *appraisal* of research programs, which can only be done with historical hindsight, and *advice* to scientists, which he denied it was the purpose of his methodology to offer. 'There is no instant rationality in science' became one of Lakatos's slogans, capturing the sense in which he considered positivism and falsificationism, insofar as they can be interpreted as offering criteria that can be used for the acceptance and rejection of theories, as striving for too much.

Problems with Lakatos's methodology

As we have seen, Lakatos regarded it as appropriate to test methodologies against the history of science. It is therefore legitimate, even in his own terms, to raise the question of whether his methodology is descriptively adequate. There are grounds for doubting that it is. For instance, are there such things as 'hard cores' serving to identify research programs to be found in the history of science? Counter evidence comes from the extent to which scientists do on occasions attempt to solve problems by adjusting the fundamentals of the theories or programs in which they work. Copernicus himself, for example, moved the sun a little to the side of the centres of planetary orbits, had the moon orbit the earth rather than the sun, and came to use all sorts of devices to adjust the details of the epicyclical motions, to the extent that those motions ceased to be uniform. So what exactly was the hard core of the Copernican program? In the nineteenth century there were serious attempts to cope with problems such as the motion of the planet Mercury by modifying the inverse square law of attraction. There are violations of some of Lakatos's own prime examples of hard cores to be found in history, therefore.

A deeper problem concerns the reality or otherwise of the methodological decisions that play such an important role in Lakatos's account of science. For instance, as we have seen, according to Lakatos (1970, p. 133) the hard core of a program is rendered unfalsifiable by 'the methodological decisions of its protagonists'. Are these decisions a historical reality or a figment of Lakatos's imagination? Lakatos does not really give any evidence for the answer that he needs, and it is not totally clear what kind of study would provide that evidence. The issue is a vital one for Lakatos, for the methodological decisions are the locus of the distinction between his own position and that of Kuhn. Both Kuhn and Lakatos agree that scientists work in a coordinated way within a framework. For Kuhn, in one of his moods at least, the question of how and why they do so is to be revealed by sociological analysis. For Lakatos this leads to an unacceptable relativism. So for him, the cohesion is brought about by methodological decisions that are *rational*. Lakatos does not provide an answer to the charge that these decisions have no historical (or contemporary) reality, nor does he give a clear answer to the question of the sense in which they should be regarded as rational.

Another fundamental criticism of Lakatos is directly connected with the central theme of this book, the question of what, if anything, is characteristic of scientific knowledge. Lakatos's rehetoric, at least, suggests that his methodology was intended to give a definitive answer to that question. He claimed that the 'central problem in the philosophy of science is – the problem of stating *universal* conditions under which a theory is scientific', a problem that is 'closely linked with the problem of the rationality of science' and whose solution 'ought to give us guidance as to when the acceptance of a scientific theory is rational or not' (Worrall and Currie, 1978a, pp. 168–9, italics in original). Lakatos (1970, p. 176) portrayed his methodology as a solution to these problems that would 'help us in devising laws for stemming – intellectual pollution'. 'I [Lakatos] give criteria for progression and stagnation within a program and also rules for the "elimination" of whole research programs' (Worrall and Currie, 1978a, p. 112). It is clear from the details of Lakatos's position, and his own comments on those details, that Lakatos's methodology was not capable of living up to these expectations. He did not give rules for the elimination of whole research programs because it is rational to stick to a degenerating program in the hope that it will make a comeback. And if it was scientific to stick to the Copernican theory for the century that it took for that theory to bear significant fruit, why aren't contemporary Marxists (one of Lakatos's prime targets) scientific in attempting to develop historical materialism to a point where it will bear significant fruit? Lakatos in effect conceded that his methodology was in no position to diagnose any contemporary theory as non-scientific 'intellectual pollution' once he recognised and acknowledged, in the context of physical science, that his methodology could only make judgments in retrospect, with the benefit of historical hindsight. If there is no 'instant rationality' then there can be no on-the-spot rejection of Marxism, sociology or any other of Lakatos's *bêtes noir.*

Another basic problem with Lakatos's methodology stems from the way in which he deemed it necessary to support it by studies from the history of science. Lakatos and his followers made the necessary case by means of case studies of physical sciences over the last three hundred years. But if the methodology supported in this way is then used to judge other areas, such as Marxism or astrology, what is in effect being assumed without argument is that all areas of study, if they are to be regarded as 'scientific', must share the basic characteristics of physics. Paul Feyerabend (1976) has criticised Lakatos in this way. Lakatos's procedure certainly begs an important fundamental question and has only to be explicitly stated to reveal a problem. There are a number of prima facie reasons at least why one might expect that a methodology and set of standards for judging physics might not be appropriate in other areas. Physics can, and often does, proceed by isolating individual mechanisms – gravity, electromagnetic forces, the mechanisms at work when fundamental particles collide and so on – in the artificial circumstances of a controlled experiment. People and societies cannot in general be treated in this way without destroying what it is that is being investigated. A great deal of complexity is necessary for living systems to function as such, so even biology can be expected to exhibit some important differences from physics. In social sciences the knowledge that is produced itself forms an important component of the systems being studied. So, for example, economic theories can affect the way in which individuals operate in the market place, so that a change in theory can bring about a change in the economic system being studied. This is a complication that does not apply in the physical sciences. The planets do not change their motions in the light of our theories about those motions. Whatever the force of the arguments that can be developed from reflections such as these, it remains the case that Lakatos presupposes, without argument, that all scientific knowledge should in some fundamental sense be like the physics of the last three hundred years.

Another fundamental issue is brought to light when we consider the implications of a study by Lakatos (1976a), published posthumously, on 'Newton's effect on scientific standards'. In that study, Lakatos makes the case that Newton, in practice, brought about a change in scientific standards, a change that Lakatos clearly regards as progressive. But the fact that Lakatos can make such a case does not rest easily with the assumption he makes repeatedly elsewhere, that an appraisal of science must be made with respect to some 'universal' criterion. If Newton changed scientific standards for the better, then one can ask, 'with respect to what standard was the change progressive?' We have a problem of a similar kind to the one that confronted Kuhn. It is a problem we will need to confront, or perhaps dispel, later in this book.

Further reading

The central text for Lakatos's methodology is his 1970 text, 'Falsification and the Methodology of Scientific Research Programmes'. Most of the other key papers have been collected in Worrall and Currie (1978a and 1978b). Also important is Lakatos (1968), *The Problem of Inductive Logic*, and (1971), 'Replies to Critics'. A fascinating account of Lakatos's application of his ideas to mathematics is his *Proofs and Refutations* (1976b). Howson (1976) contains historical case studies designed to support Lakatos's position. Another such study is Lakatos and Zahar (1975). Cohen, Feyerabend and Wartofsky (1976) is a collection of essays in memory of Lakatos. Feyerabend (1976) is an important critique of Lakatos's methodology. The notion of a novel prediction is discussed by Musgrave (1974b), Worrall (1985), Worrall (1989a) and Mayo (1996). A useful overview of Lakatos's work is B. Larvor (1998), *Lakatos: An Introduction*.

CHAPTER 10

Feyerabend's anarchistic theory of science

The story so far

We seem to be having trouble with our search for *the* characterisation of science that will serve to pick out what distinguishes it from other kinds of knowledge. We started with the idea, adopted by the positivists who were so influential earlier in the twentieth century, that science is special because it is derived from the facts, but this attempt floundered because facts are not sufficiently straightforward for this view to be sustained, since they are 'theory-dependent' and fallible, and because no clear account of how theories can be 'derived' from the facts could be found. Falsificationism did not fare much better, mainly because in any realistic situation in science it is not possible to locate the cause of a faulty prediction, so a clear sense of how theories can be falsified becomes almost as elusive as a clear sense of how they can be confirmed. Both Kuhn and Lakatos tried to solve the problem by focusing attention on the theoretical framework in which scientists work. However, Kuhn, for his part, stressed the extent to which workers in rival paradigms 'live in different worlds' to such a degree that he left himself with inadequate resources for elucidating a sense in which a change from one paradigm to another in the course of a scientific revolution is a step forward. Lakatos tried to avoid that trap, but, apart from problems concerning the reality of the methodological decisions he freely invoked in his answer, he ended up with a criterion for characterising science that was so lax that few intellectual pursuits could be ruled out. One philosopher of science who was not surprised by, and who attempted to draw out what he saw to be the full implications of, these failures was Paul Feyerabend, whose controversial but nevertheless influential 'anarchistic' account of science is described and assessed in this chapter.

Feyerabend's case against method

Paul Feyerabend, an Austrian who was based in Berkeley, California, for most of his academic career, but who also spent time interacting with (and antagonising) Popper and Lakatos in London, published a book in 1975 with the title *Against Method: Outline of an Anarchistic Theory of Knowledge*. In it he challenged all of the attempts to give an account of scientific method that would serve to capture its special status by arguing that there is no such method and, indeed, that science does not possess features that render it necessarily superior to other forms of knowledge. If there is a single, unchanging principle of scientific method, Feyerabend came to profess, it is the principle 'anything goes'. There are passages in Feyerabend's writings, both early and late, that can be drawn on to severely qualify the extreme anarchistic account of science that is contained in the bulk of *Against Method*. However, it will be most instructive for our purpose to stick to the unqualified, anarchistic theory of science to see what we can learn from it. In any case, it is the extreme form of Feyerabend's position that has made its mark in the literature and which philosophers of science have, not without difficulty, attempted to counter.

Feyerabend's main line of argument attempts to undermine characterisations of method and progress in science offered by philosophers by challenging them on their own ground in the following way. He takes examples of scientific change which his opponents (including the vast majority of philosophers) consider to be classic instances of scientific progress and shows that, as a matter of historical fact, those changes did not conform to the theories of science proposed by those philosophers. (Feyerabend does not have to himself agree that the episodes in question were progressive for his argument to go through.) The main example appealed to by Feyerabend involves the advances in physics and astronomy made by Galileo. Feyerabend's point is that if an account of method and progress in science cannot even make sense of Galileo's innovations, then it is not much of an account of science. In this outline of Feyerabend's position I will stick largely to the Galileo example, mainly because it is sufficient to illustrate Feyerabend's position, but also because the example is readily understood without requiring resort to recondite technicalities.

A number of Feyerabend's points will be familiar because I have already drawn on them for various purposes earlier in this book.

Quotations invoked in chapter 1 of this book illustrate the positivist or inductivist view that Galileo's innovations can be explained in terms of the extent to which he took the observable facts seriously and built his theories to fit them. The following passage from Galileo's *Dialogue Concerning the Two Chief World Systems* (1967), cited by Feyerabend (1975, pp. 100–1), indicates that Galileo thought otherwise.

> You wonder that there are so few followers of the Pythagorean opinion [that the earth moves] while I am astonished that there have been any up to this day who have embraced and followed it. Nor can I ever sufficiently admire the outstanding acumen of those who have taken hold of this opinion and accepted it as true: they have, through sheer force of intellect done such violence to their own senses as to prefer what reason told them over that which sensible experience plainly showed them to the contrary. For the arguments against the whirling of the earth we have already examined are very plausible, as we have seen: and the fact that the Ptolemaics and the Aristotelians and all their disciples took them to be conclusive is indeed a strong argument of their effectiveness. But the experiences which overtly contradict the annual movement are indeed so much greater in their apparent force that, I repeat, there is no limit to my astonishment when I reflect that Aristarchus and Copernicus were able to make reason so conquer sense that, in defiance of the latter, the former became mistress of their belief.

Far from accepting the facts considered to be borne out by the senses by his contemporaries, it was necessary for Galileo (1967, p. 328) to conquer sense by reason and even to replace the senses by 'a superior and better sense', namely, the telescope. Let us consider two instances where Galileo needed to 'conquer' the evidence of the senses – his rejection of the claim that the earth is stationary and his rejection of the claim that the apparent sizes of Venus and Mars do not change appreciably during the course of the year.

If a stone is dropped from the top of a tower it falls to the base of the tower. This, and other experiences like it, can be taken as evidence that the earth is stationary. For if the earth moves, spinning on its axis, say (the whirling of the earth referred to by Galileo in the passage cited), then should it not move from beneath the stone during its fall, with the result that the stone should fall some distance from the base of the tower? Did Galileo reject this argument by appealing to the facts? That is certainly not how Galileo did it in the *Dialogue*, as Feyerabend pointed out. Galileo (1967, p. 125 ff) achieved the desired result by 'picking the brains' of the reader. He argued as follows. The speed of a ball set rolling down a frictionless slope will increase, because it is 'falling' towards the centre of the earth to some degree. Conversely, the speed of a ball rolled up a frictionless slope will decrease because it is rising away from the centre of the earth. Having persuaded the reader to accept this as obvious, he or she is now asked what will happen to the speed of the ball if the slope is perfectly horizontal. It would seem that the answer is that the speed will neither increase nor decrease since the ball will be neither rising nor falling. The horizontal motion of the ball persists and remains constant. Although this falls short of Newton's law of inertia, it is an example of a uniform motion that persists without a cause, and it is sufficient for Galileo to counter a range of arguments against the spinning earth. Galileo draws the implication that the horizontal motion of the stone falling from the tower, which it shares with the tower as the earth spins, remains unchanged. That is why it stays with the tower, striking the ground at its foot. So the tower argument does not establish that the earth is stationary in the way many had supposed. To the extent that Galileo's case was successful it did not involve appealing to the results of observation and experiment, at his own admittance. (I point out here that frictionless slopes were even harder to obtain in Galileo's time than they are now, and that measuring the speed of a ball at various locations on the slope lay beyond what was feasible at the time.)

We saw in chapter 1 that the apparent sizes of Venus and Mars were important insofar as the Copernican theory predicted that they should change appreciably, a prediction not borne out by naked-eye observations. The problem is resolved once the telescopic rather than the naked-eye data is accepted. But how was the preference for the telescopic data to be defended? Feyerabend's rendering of the situation and Galileo's response to it run as follows. Accepting what the telescope revealed in the astronomical context was by no means straightforward. Galileo did not have an adequate or detailed theory of the telescope, so he could not defend the telescopic data by appeal to one. It is true that in a terrestrial context there were trial and error methods of vindicating telescopic sightings. For instance, the reading of an inscription on a distant building, indiscernible to the naked eye, could be checked by going close to the building, and the identification of the cargo of a distant ship could be vindicated once the ship arrived in port. But the vindication of terrestrial use could not be straightforwardly employed to justify astronomical use of the telescope. Terrestrial use of the telescope is aided by a range of visual cues absent in the astronomical case. Genuine images can be distinguished from many artifacts of the telescope because we are familiar with the kinds of things being inspected. So, for instance, if the telescope reveals the mast of a distant ship to be wavy, red on one side and blue on the other and accompanied by black specks hovering above it, the distortions, colours and specks can be dismissed as artifacts. However, when looking into the heavens, we are in unfamiliar territory and lack clear guidance as to what is really there as opposed to an artifact. What is more, comparison with familiar objects to help judge size, and the use of parallax and overlap to help judge what is far and what is near, is a luxury not in general available in astronomy and it is certainly not the case that Galileo could check telescopic sightings of planets by moving closer to them to check with the naked eye. There was even direct evidence that the telescopic data was erratic insofar as it magnified the moon to a different degree than it magnified the planets and stars.

According to Feyerabend (1975, p. 141), these difficulties were such that recourse to argument would have been inadequate for the task of convincing those opponents who wished to deny both the Copernican theory and the telescopic data relating to the heavens. Consequently, Galileo needed to, and did, resort to propaganda and trickery.

> On the other hand, there are some telescopic phenomena which are plainly Copernican. Galileo introduces these phenomena as independent evidence for Copernicus while the situation is rather that one refuted view – Copernicanism – has a certain similarity to phenomena emerging from another refuted view – the idea that telescopic phenomena are faithful images of the sky. Galileo prevails because of his style and his clever techniques of persuasion, because he writes in Italian rather than in Latin, and because he appeals to people who are temperamentally opposed to the old ideas and the standards of learning connected with them.

It should be clear that if Feyerabend's construal of Galileo's methodology is correct and typical of science, then standard positivist, inductivist and falsificationist accounts of science have serious problems accommodating it. It can be accommodated into Lakatos's methodology, according to Feyerabend, but only because that methodology is so lax that it can accommodate almost anything. Feyerabend teased Lakatos by welcoming him as a 'fellow anarchist', albeit one 'in disguise', playfully dedicating *Against Method* to Lakatos 'friend and fellow anarchist'. The way in which Feyerabend construes the two frameworks, the Aristotelian stationary earth framework backed up by naked-eye data and the Copernican, moving earth theory supported by telescopic data, as mutually exclusive circles of thought, as it were, is reminiscent of Kuhn's portrayal of paradigms as mutually exclusive ways of seeing the world. Indeed, the two philosophers both independently coined the word 'incommensurable' to describe the relationship between two theories or paradigms that cannot be logically compared for lack of theory-neutral facts to exploit in the comparison. Kuhn avoided Feyerabend's anarchistic conclusions essentially by appealing to social consensus to restore law and order. Feyerabend (1970) rejected Kuhn's appeal to the social consensus of the scientific community, partly because he did not think Kuhn distinguished between legitimate and illegitimate ways (for example by killing all opponents) of achieving consensus, and also because he did not think the appeal to consensus was capable of distinguishing between science and other activities such as theology and organised crime.

Given the failure of attempts to capture the special features of scientific knowledge that render it superior to other forms, which failure Feyerabend considered himself to have established, he drew the conclusion that the high status attributed to science in our society, and the superiority it is presumed to have not only over Marxism, say, but over such things as black magic and voodoo, are not justified. According to Feyerabend, the high regard for science is a dangerous dogma, playing a repressive role similar to that which he portrays Christianity as having played in the seventeenth century, having in mind such things as Galileo's struggles with the Church.

Feyerabend's advocacy of freedom

Feyerabend's theory of science is situated in an ethical framework that places a high value on individual freedom, involving an attitude that Feyerabend described as the 'humanitarian attitude'. According to that attitude, individual humans should be free and possess liberty in something like the sense the nineteenth-century philosopher John Stuart Mill (1975) defended in his essay 'On Liberty'. Feyerabend (1975, p. 20) declared himself in favour of 'the attempt to increase liberty, to lead a full and rewarding life' and supports Mill in advocating 'the cultivation of individuality which alone produces, or can produce, well-developed human beings'. From this humanitarian point of view, Feyerabend supports his anarchistic account of science on the grounds that it increases the freedom of scientists by removing them from methodological constraints and, more generally, leaves individuals the freedom to choose between science and other forms of knowledge.

From Feyerabend's point of view, the institutionalisation of science in our society is inconsistent with the humanitarian attitude. In schools, for example, science is taught as a matter of course. 'Thus, while an American can now choose the religion he likes, he is still not permitted to demand that his children learn magic rather than science at school. There is a separation between state and Church, there is no separation between state and science' (1975, p. 299). What we need to do in the light of this, wrote Feyerabend (1975, p. 307), is to 'free society from the strangling hold of an ideologically petrified science just as our ancestors freed *us* from the strangling hold of the One True Religion!' In Feyerabend's image of a free society, science will not be given preference over other forms of knowledge or over other traditions. A mature citizen in a free society is 'a person who has learned to make up his mind and who has then *decided* in favour of what he thinks suits him best'. Science will be studied as a historical phenomenon 'together with other fairy tales such as the myths of "primitive" societies' so that each individual 'has the information needed for arriving at a free decision' (1975, p. 308, italics in original). In Feyerabend's ideal society the state is ideologically neutral between ideologies to ensure that individuals maintain freedom of choice and do not have an ideology imposed on them against their will.

The culmination of Feyerabend's case against method, together with his advocacy of a particular brand of freedom for the individual, is his anarchistic theory of knowledge (1975, pp. 284–5, italics in original).

> None of the methods which Carnap, Hempel, Nagel [three prominent positivists], Popper or even Lakatos want to use for rationalising scientific changes can be applied, and the one that can be applied, refutation, is greatly reduced in strength. What remains are aesthetic judgments, judgments of taste, metaphysical prejudices, religious desires, in short, *what remains are our subjective wishes*: science at its most advanced and general returns to the individual a freedom he seems to lose in its more pedestrian parts.

There is no scientific method, then. Scientists should follow their subjective wishes. Anything goes.

Critique of Feyerabend's individualism

A critique of Feyerabend's understanding of human freedom will act as a useful preliminary to an appraisal of his critique of method. A central problem with Feyerabend's notion of freedom stems from the degree to which it is entirely negative, in the sense that freedom is understood as freedom from constraints. Individuals should be free of constraints to the extent that they can follow their subjective wishes and do what they like. This overlooks the positive side of the issue, the extent to which individuals have access to the means to fulfil their wishes. For example, freedom of speech can be, and often is, discussed in terms of freedom from constraints, in the form of state suppression, libel laws and the like. So, for example, if students disrupt a lecture on campus by an academic expressing views sympathetic to Fascism they might well be accused of denying the speaker freedom of speech. They are accused of putting an obstacle in the way of the speaker's natural right. However, freedom of speech can be considered, from the positive point of view, in terms of the resources available to individuals to have their views heard by others. What access does a particular individual have to the media, for example? This point of view puts our example in a different light. The disruption of the lecture could perhaps be justified on the grounds that the speaker was given access to a university lecture hall, microphone, media advertising and so on in a way that those advocating other views were not. The eighteenth-century philosopher David Hume nicely illustrated the point I am getting at when he criticised John Locke's idea of the Social Contract. Locke had construed the social contract as being freely adopted by members of a democratic society and argued that anyone not wishing to subscribe to the contract was free to emigrate. Hume responded as follows:

> Can we seriously say, that a poor peasant or artisan has a free choice to leave his country, when he knows no foreign language or manners, and lives from day to day, by the small wages which he acquires? We may as well assert that a man, by remaining in a vessel, freely consents to the domination of the master; though he was carried on board while asleep, and must leap into the ocean and perish, the moment he leaves her.[1]

Individuals are born into a society that pre-exists them and which, in that sense, possesses characteristics they do not choose and cannot be in a position to choose. The courses of action open to them, and, consequently, the precise senses in which they are free, will be determined by the access that they have in practice to the resources necessary for various courses of action. In science too an individual who wishes to make a contribution to a science will be confronted by the situation as it stands: various theories, mathematical techniques, instruments and experimental techniques. The paths of action open to scientists in general will be delimited by that objectively existing situation, while the paths open to a particular scientist will be determined by the subset of the existing resources to which that individual scientist has access. Scientists will be free to follow their 'subjective wishes' only insofar as they are free to choose among the restricted range of options open to them. What is more, a prerequisite for an understanding of that situation will be a characterisation of the situation that individuals face, like it or not. Whether it be changes in science or in society generally, the main theoretical work involves understanding the situations confronted by individuals rather than involving some generalised appeal to unconstrained freedom.

It is ironic that Feyerabend, who in his study of science goes to great lengths to deny the existence of theory-neutral facts, in his social theory appeals to the far more ambitious notion of an ideology-neutral state. How on earth would such a state come into existence, how would it function and what would sustain it? In the light of work that has been done in making serious attempts to get to grips with questions about the origin and nature of 'the state', Feyerabend's fanciful speculations about a utopia in which all individuals are free to follow their inclinations in an unrestricted way appear childish.

Criticising Feyerabend for setting his views on science in an individualist framework involving a naive notion of freedom is one thing. Getting to grips with the details of the case he makes 'against method' in science is another. In the next chapter we will see what can be constructively salvaged from Feyerabend's attack on method.

Further reading

Feyerabend develops some of the ideas of his *Against Method: Outline of an Anarchistic Theory of Knowledge* (1975) in *Science in a Free*

Society (1978). *Realism, Rationalism and Scientific Method* (Feyerabend, 1981a) and *Problems of Empiricism* (Feyerabend, 1981b) are collections of his articles, a number of which predate his 'anarchistic' phase. 'Consolations for the Specialist' (1970) and 'On the Critique of Scientific Reason' (1976) are his critiques of Kuhn and Lakatos respectively. I have taken issue with Feyerabend's portrayal of Galileo's science in 'Galileo's Telescopic Observations of Venus and Mars' (Chalmers, 1985) and 'The Galileo that Feyerabend Missed' (Chalmers, 1986). Books on Feyerabend's philosophy of science include Couvalis (1989), Farrell (2003) and Oberheim (2006).

CHAPTER 11

Methodical changes in method

Against universal method

We saw in the previous chapter that Feyerabend made a case against the various accounts of scientific method that have been put forward by philosophers as attempts to capture the distinctive feature of scientific knowledge. A key strategy that he employed was to argue for the incompatibility of those accounts and Galileo's advances in physics and astronomy. Elsewhere (in Chalmers, 1985 and 1986) I have taken issue with Feyerabend's historical account of the Galileo episode and some of the details of my disagreement will be introduced and exploited in the next section. Once that history is corrected I believe it to remain the case that the corrected history poses problems for standard accounts of science and the scientific method. That is, I suggest there is a sense in which Feyerabend's case against method can be sustained, *provided we are clear about the notion of method that has been refuted.* Feyerabend's case tells against the claim that there is a universal, ahistorical method of science that contains standards that all sciences should live up to if they are to be worthy of the title 'science'. Here the term 'universal' is used to indicate that the proposed method is to apply to all sciences or putative sciences – physics, psychology, creation science or whatever – while the term 'ahistorical' signals the timeless character of the method. It is to be used to appraise Aristotle's physics as much as Einstein's and Democritus's atomism as much as modern atomic physics. I am happy to join Feyerabend in regarding the idea of a universal and ahistoric method as highly implausible and even absurd. As Feyerabend (1975, p. 295) says, 'The idea that science can, and should, be run according to fixed and universal rules is both unrealistic and pernicious', is 'detrimental to science, for it neglects the complex physical and historical conditions which influence scientific change' and 'makes science less adaptable and more dogmatic'. If there is to be a scientific method capable of judging sciences of all kinds, past, present *and future*, one might well ask what resources philosophers have for arriving at such a potent tool, so potent that it can tell us in advance what are the appropriate standards for judging future science. If we have a conception of science as an open-ended quest to improve our knowledge, then why cannot there be room for us to *improve* our methods and adapt and refine our standards in the light of what we learn?

I have no problem joining the campaign that Feyerabend launched against method, then, provided method is understood as universal, unchanging method. We have seen that Feyerabend's response to the case against method is to assume that there is no method, that scientists should follow their own subjective wishes and that anything goes. However, universal method and no method at all do not exhaust the range of possibilities. A middle way would hold that there are methods and standards in science, but that they can vary from science to science and can, within a science, be changed, and changed for the better. Not only does Feyerabend's case not tell against this intermediate view, but his Galileo example can be construed in a way that supports it, as I shall attempt to show in the next section.

I hold that there is a middle way, according to which there are historically contingent methods and standards implicit in successful sciences. A common response from philosophers of science who reject Feyerabend's anarchism and extreme relativism as firmly as I do is that those like myself who seek a middle way are kidding ourselves. John Worrall (1988), for instance, has given clear expression to the general line of argument. If I am to defend a change in scientific method in a way that avoids extreme relativism then I am obliged to show in what way such a change is for the better. But better according to what standards? It would seem that unless there are some superstandards for judging changes in standards then those changes cannot be construed in a non-relativist way. But superstandards take us back to the universal method that is meant to yield such standards. So, Worrall's argument goes, either we have universal method or relativism. There is no middle way. As at least a preliminary to a rejoinder to this argument it is useful to take an example from science of a change in standards. The next section is devoted to such a change accomplished by Galileo.

Telescopic for naked-eye data: a change in standards

One of Galileo's Aristotelian opponents (cited in Galileo, 1967, p. 248) referred to the idea that 'the senses and experience should be our guide in philosophising' as 'the criterion of science itself'. A number of commentators on the Aristotelian tradition have noted that it was a key principle within that tradition that knowledge claims should be compatible with the evidence of the senses when they are used with sufficient care under suitable conditions. Ludovico Geymonat (1965, p. 45), a biographer of Galileo, refers to the belief 'shared by most scholars at the time [of Galileo's innovations]' that 'only direct vision has the power to grasp actual reality'. Maurice Clavelin (1974, p. 384), in a context where he is comparing Galilean and Aristotelian science, observes that 'the chief maxim of Peripatetic physics was never to oppose the evidence of the senses', and Stephen Gaukroger (1978, p. 92), in a similar context, writes of 'a fundamental and exclusive reliance on sense-perception in Aristotle's works'. A teleological defence of this fundamental standard was common. The *function* of the senses was understood to be to provide us with information about the world. Therefore, although the senses can mislead in abnormal circumstances, for instance in a mist or when the observer is sick or drunk, it makes no sense to assume that the senses can be systematically misleading when they are fulfilling the task for which they are intended. Irving Block (1961, p. 9), in an illuminating article on Aristotle's theory of sense perception, characterises Aristotle's view as follows:

> Nature made everything for a purpose, and the purpose of man is to understand Nature through science. Thus it would have been a contradiction for Nature to have fashioned man and his organs in such a way that all knowledge and science must, from its inception, be false.

Aristotle's views were echoed by Thomas Aquinas many centuries later, as Block (1961, p.7) reports:

> Sense perception is always truthful with respect to its proper objects, for natural powers do not, as a general rule, fail in the activities proper to them, and if they do fail, this is due to some derangement or other. Thus, only in a minority of cases do the senses judge inaccurately of their proper objects, and then only through some organic defect, e.g. when people sick with fever taste sweet things as bitter because their tongues are ill-disposed.

Galileo was faced with a situation in which a reliance on the senses, including naked-eye data was 'a criterion of science itself'. In order to introduce the telescope, and have telescopic data replace and overrule some naked-eye data, he needed to fly in the face of this criterion. By the time he had done so, he had effected a change in the standards of science. As we have seen, Feyerabend did not believe it was possible for Galileo to make a compelling case and needed to resort to propaganda and trickery. The historical facts tell otherwise.

I have already considered (pp. 20–3) the case that Galileo made for the veracity of his sightings of the moons of Jupiter. Here I will focus on the case that Galileo was able to muster for accepting what the telescope revealed of the changing apparent sizes of Venus and Mars. We have already described, in the previous chapter, the urgency of the question and also accepted Feyerabend's account of the difficulties that lay in the way of accepting telescopic observations of the heavens.

Galileo appealed to the phenomenon of irradiation to help discredit naked-eye observations of the planets and as providing grounds for preferring the telescopic observations. Galileo's hypothesis (1967, p. 333) was that the eye 'introduces a hindrance of its own' when it views small, bright, distant light sources against a dark background. Because of this, such objects appear 'festooned with adventitious and alien rays'. Thus, Galileo (1957, p. 46) explained elsewhere, if stars 'are viewed by means of unaided vision, they present themselves to us not as of their simple (and, so to speak, their physical) size but as irradiated by a certain fulgor and as fringed with sparkling rays'. In the case of the planets irradiation is removed by the telescope.

Since Galileo's hypothesis involves the claim that irradiation arises as a consequence of the brightness, smallness and distance of the source, it can be tested by modifying those factors in a variety of ways that do not involve use of the telescope. A number of ways are explicitly invoked by Galileo (1957, pp. 46–7). The brightness of stars and planets can be reduced by viewing them through a cloud, a black veil, coloured glass, a tube, a gap between the fingers or a pinhole in a card. In the case of planets the irradiation is removed by these techniques, so that they 'show their globes perfectly round and definitely bounded', whereas in the case of stars the irradiation is never completely removed, so that they are 'never seen to be bounded by a circular periphery, but have rather the aspect of blazes whose rays vibrate about them and scintillate a great deal'. As far as the dependence of irradiation on the apparent size of the observed light source is concerned, Galileo's hypothesis is borne out by the fact that the moon and the sun are not subject to irradiation. This aspect of Galileo's hypothesis, as well as the associated dependence of irradiation on the distance of the source, can be subject to a direct terrestrial test. A lighted torch can be viewed from near or far and at day or night. When viewed at a distance at night, when it is bright compared with its surroundings, it appears larger than its true size. Accordingly, Galileo (1967, p. 361) remarked that his predecessors, including Tycho and Clavius, should have proceeded with more caution when estimating the size of stars.

> I will not believe that they thought that the true disc of a torch was as it appears in profound darkness, rather than as it is when perceived in lighted surroundings: for our lights seen from afar at night look large, but from near at hand their true flames are seen to be small and circumscribed.

The dependence of irradiation on the brightness of a source relative to its surroundings is further confirmed by the appearance of stars at twilight, which appear much smaller then than at night, and of Venus when observed in broad daylight, which appears 'so small that it takes sharp eyesight to see it, though in the following night it appears like a great torch'. This latter effect provides a rough way of testing for the predicted change in size of Venus that does not involve an appeal to telescopic evidence. The test can be made with the naked eye provided observations are restricted to daytime or twilight. According to Galileo, at least, the changes in size are 'quite perceptible to the naked eye', although they can only be observed precisely with the telescope (Drake, 1957, p. 131).

By fairly straightforward practical demonstration, then, Galileo was able to show that the naked eye yields inconsistent information when small light sources, bright compared with their surroundings, are viewed in the terrestrial and celestial domain. The phenomenon of irradiation, for which Galileo provided a range of evidence, as well as the more direct demonstration with the lamp, indicated that naked-eye observations of small, bright light sources are unreliable. One implication of this is that naked-eye observations of Venus in daylight are to be preferred to those made at night when Venus is bright compared with its surroundings. The former, unlike the latter, show that the apparent size of Venus varies during the course of the year. All this can be said without any reference to the telescope. When we now note that the telescope removes irradiation when used to observe planets and that, what is more, the variations in apparent size are compatible with the variations observable with the naked eye in daylight, a strong case for the telescopic data begins to emerge.

A final argument for the veracity of the telescopic data on the sizes of Venus and Mars is that they corresponded precisely with the predictions of all of the serious astronomical theories at the time. This conflicts with the way in which Feyerabend, and Galileo himself, presented the situation, implying, as they did, that the data offers support to the Copernican theory over its rivals. The rivals to the Copernican theory were those of Ptolemy and Tycho Brahé. Both of those theories predicted precisely the same variations in size as the Copernican theory did. Variations in distance from earth, leading to predicted changes in apparent size, arise in the Ptolemaic system because the planets move closer then further from the earth as they traverse the epicycles superimposed on the deferents, which latter were equidistant from the earth. They occur in Tycho Brahé's system, in which planets other than earth orbit the sun while the sun itself orbits a stationary earth, for the same reason that they occur in the Copernican theory, since the two are geometrically equivalent. Derek J. de S. Price (1969) has shown quite generally that this must be so once the systems are adjusted to fit the observed angular positions of the planets and the sun. That the apparent sizes of the planets had posed a problem for the major astronomical theories since antiquity is acknowledged by Osiander in his introduction to Copernicus's *Revolutions of the Heavenly Spheres.*

We have surveyed the way in which Galileo argued for acceptance of some significant telescopic findings, arguments that, I suggest, were compelling, a suggestion borne out by the historical fact that they convinced all of Galileo's serious rivals in a short space of time. But in establishing his case, Galileo made the first step in what was to be a common trend in science, the replacement of naked-eye data by data acquired by way of instruments, and in doing so violated, and brought about a change in, 'the criterion of science itself'. How does his accomplishment of this bear on the case for and against method?

Piecemeal change of theory, method and standards

How is it that Galileo has managed to change standards by making a rational case in the face of arguments, such as John Worrall's, to the effect that this is impossible? He was able to do so because there was much that was shared between him and his rivals. There was a large overlap in what they aimed for. Among much else, they shared the aim of giving a description of the motions of the heavenly bodies that was borne out by the empirical evidence. After all, Ptolemy's *Almagest* is full of recordings of planetary positions, and Tycho Brahé is famed for his construction of massive quadrants and the like, which dramatically increased the accuracy of such recordings. There were low-level observations pointed out by Galileo that his opponents had no sensible option but to accept, such as the observation that a lamp appears larger than it really is from a distance at night, and that Venus looks smaller in the light of day than in the dark of night. Shared observations such as these, against the background of the shared aim, were sufficient for Galileo to be able to convince his opponents, using 'clever techniques of persuasion' that involved nothing other than straightforward argument, that in one context at least they should be willing to abandon the 'criterion of science itself' and accept some telescopic data rather than their naked-eye counterpart.

At any stage in its development, a science will consist of some specific aims to arrive at knowledge of some specified kind, methods for arriving at those aims together with standards for judging the extent to which they have been met, and specific facts and theories that represent the current state of play as far as the realisation of the aim is concerned. Each individual item in the web of entities will be subject to revision in the light of research. We have already discussed ways in which theories and facts are fallible (remember that supercooled liquids refute the claim that liquids cannot flow uphill) and we illustrated in the previous section a change in method and standards. The detailed form that the aim of a science takes can change too. Let me give an example.

The experimental work of Robert Boyle is rightly seen as a major contribution to the scientific revolution of the seventeenth century. Two somewhat conflicting aspects of Boyle's work can be discerned that, in a sense, represent the old and the new way of doing science. In his more philosophical writings Boyle advocated the 'mechanical philosophy'. According to that philosophy, the material world is seen as consisting of pieces of matter. It is taken as obvious that there is just this one kind of matter. Observable-sized objects are made up of arrangements of microscopic corpuscles of matter, and change is to be understood in terms of the rearrangement of corpuscles. The only properties corpuscles of matter have are the specific size, shape and motion that each one possesses, together with the property of impenetrability that serves to distinguish matter from empty space. The motion of a corpuscle changes when it collides with another, and this mechanism is the source of all activity and change in nature. An explanation of some physical process will involve tracing that process back to the motions, collisions and rearrangements of the corpuscles involved. In giving expression to a version of this view, Boyle was subscribing to the new mechanical world view that was seen as the appropriate alternative to the Aristotelian one. In it, adequate explanations were ultimate explanations. They appealed to the shapes, sizes, motions and collisions of corpuscles, and these notions were themselves not considered to be in need of explanation. The aim of science, then, from this point of view, is ultimate explanations.

As well as advocating the mechanical philosophy, Boyle did experiments, notably his experiments in pneumatics and chemistry. As some of Boyle's own remarks imply, his experimental successes did not yield scientific knowledge of the kind demanded within the mechanical philosophy. Boyle's experiments on the physics of air, especially those with an air pump which enabled him to evacuate most of the air from a glass chamber, led him to explain a range of phenomena, such as the behaviour of barometers both inside and outside of evacuated chambers, in terms of the weight and elasticity of air. He was even able to suggest a version of the law connecting the pressure and volume of a fixed mass of gas that bears his name. But his explanations were not scientific explanations from the point of view of the mechanical philosophy because they were not ultimate. Appealing to weight and elasticity was not acceptable until those properties themselves had been explained in terms of corpuscular mechanisms. Needless to say, Boyle was unable to satisfy that demand. Eventually it became appreciated that Boyle's experimental science sought explanations that were both useful and attainable. By contrast, mechanical explanations in the strict sense came to be appreciated as unattainable. In effect, by the end of the seventeenth century the aim for ultimate explanations was given up in physics. That aim came to be seen as utopian, especially when contrasted with the achievements of experimental science.

The general idea, then, is that any part of the web of aims, methods, standards, theories and observational facts that constitute a science at a particular time can be progressively changed, and the remaining part of the web will provide the background against which a case for the change can be made. However, it will certainly not be possible to make a reasoned case for changing everything in the web at once, for then there would be no ground on which to stand to make such a case. So if it were typical of science that rival scientists see everything differently from the point of view of their respective paradigms and live in different worlds to the extent that they share nothing, it would indeed be impossible to capture an objective sense in which science progresses. But there are no situations in science or its history or, for that matter, anywhere else that conform to this caricature. We do not need a universal, ahistorical account of scientific method to give an objective account of progress in science, and, furthermore, an objective account of how method can be changed for the better is possible.

A light-hearted interlude

I can imagine how John Worrall, and like-minded opponents of relativism and defenders of universal method, would respond to the line I have taken above. They will say of my Galileo example, for instance, that, although it does illustrate a change in standards, an appeal to some higher, more general standards is involved. Both Galileo and his rivals demanded that their account of planetary orbits should be borne out by appropriate evidence, for example. Once we have spelt out these general assumptions, my critics might well argue, then it is those general assumptions that constitute universal method, and it is precisely those which form the backdrop against which the change brought about by Galileo is to be judged progressive. Without such a backdrop, I hear them say, you cannot argue that the change is progressive.

Let me make a concession. Suppose we do try to formulate some general principles that any proponent of science from Aristotle to Stephen Hawking might be expected to adhere to. Suppose the result is something like 'take argument and the available evidence seriously and do not aim for a kind of knowledge or a level of confirmation that is beyond the reach of available methods'. Let us call it the commonsense version of scientific method. I concede that there is a universal method in the commonsense. But let me immediately attempt to remove any feeling of smugness John Worrall and his allies might be enjoying having won this concession from me. Let me first point out that, to the extent that commonsense universal method is correct and adequate, it puts them all, and myself, out of business, because it is hardly the kind of thing that it takes a professional philosopher to formulate, appreciate or defend. More seriously, I point out that once we do press the issue further, and demand that more detail be given, concerning what counts as evidence and confirmation, and precisely what kind of claims can be defended and how, then those details will vary from science to science and from historical context to historical context.

A formulation of commonsense method might not be sufficiently demanding a task to keep philosophers of science in business. However, I do suggest that an appreciation of it is sufficient to resist some contemporary trends in science studies. I have in mind those sociologists of science and postmodernists (let's call them 'the levellers' for short) who downplay or deny the special status to be accorded scientific knowledge on the grounds that establishing its credentials necessarily involves the interests of scientists and groups of scientists, such things as financial or social status, professional interests and the like, in much the same way as any other social task does. In response to this I suggest there is a commonsense distinction between, say, the aim to improve knowledge of how chemicals combine and the aim to improve the social standing of professional chemists. I would even go so far as to suggest that if there are academic movements that fly in the face of this commonsense, then those in possession of such sense should demand that those movements be starved of funds. It is interesting to note that traditional philosophers of science have themselves contributed to the manufacturing of a situation that opens a space for the levellers. It is they who have presumed that a distinction between science and other kinds of knowledge can only be achieved with the aid of some philosophically articulated account of universal method. Consequently, when those attempts fail, in a way that the preceding chapters of this book have shown them to have done, the way seems open for the levellers to move in. Michael Mulkay (1979), one of the most modest of levellers to be sure, provides just one of the many possible examples of an analyst of science who draws the conclusion that a sociological categorisation of science is made necessary by the failure of what he terms 'the standard view'.[1]

This brings us to the point at which the debate within philosophy of science stood about fifteen years ago. We cannot leave matters here, because during that period there have been two important movements that have developed since then and which warrant attention. One of these movements involves an attempt to develop an account of universal method by adapting a version of probability theory. We investigate it in the next chapter. The second movement has attempted to counter what it sees as the excesses of the theory-dominated accounts of science that have held sway for some time by taking a close look at experiment and what it involves. This approach is discussed in chapter 13.

Further reading

My case against universal method is made in a little more detail in *Science and Its Fabrication* (Chalmers, 1990, chapter 2), while 'Galileo's Telescopic Observations of Venus and Mars' (Chalmers, 1985) and 'The Galileo that Feyerabend Missed' (Chalmers, 1986) contain a critique and improvement of Feyerabend's Galileo case study. Laudan (1977) and Laudan (1984) involve an attempt to find a middle way between universal method and anarchism that differs from mine. More details of the case I make with relation to Boyle's work can be found in 'The Lack of Excellence of Boyle's Mechanical Philosophy' (Chalmers, 1993) and 'Ultimate Explanation in Science' (Chalmers, 1995).

CHAPTER 12

The Bayesian approach

Introduction

Many of us had sufficient confidence in the prediction of the most recent return of Halley's comet that we booked weekends in the country, far from city lights and well in advance, in order to observe it. Our confidence proved not to be misplaced. Scientists have enough confidence in the reliability of their theories to send manned spacecraft into space. When things went amiss in one of them, we were impressed, but perhaps not surprised, when the scientists, aided by computers, were able to rapidly calculate how the remaining rocket fuel could be utilised to fire the rocket motor in just the right way to put the craft into an orbit that would return it to earth. These stories suggest that perhaps the extent to which theories are fallible, stressed by the philosophers in our story so far, from Popper to Feyerabend, are misplaced or exaggerated. Can the Popperian claim that the probability of all scientific theories is zero be reconciled with them? It is worth stressing, in this connection, that the theory used by the scientists in both of my stories was Newtonian theory, a theory falsified in a number of ways at the beginning of the twentieth century according to the Popperian account (and most others). Surely something has gone seriously wrong.

One group of philosophers who do think that something has gone radically wrong, and whose attempts to put it right have become popular in the last few decades, are the Bayesians, so called because they base their views on a theorem in probability theory proved by the eighteenth-century mathematician Thomas Bayes. The Bayesians regard it as inappropriate to ascribe zero probability to a well-confirmed theory, and they seek some kind of inductive inference that will yield non-zero probabilities for them in a way that avoids the difficulties of the kind described in chapter 4. For example, they would like to be able to show how and why a high probability can be attributed to Newtonian theory when used to calculate the orbit of Halley's comet or a spacecraft. An outline and critical appraisal of their viewpoint is given in this chapter.

Bayes' theorem

Bayes' theorem is about conditional probabilities, probabilities for propositions that depend on (and hence are conditional on) the evidence bearing on those propositions. For instance, the probabilities ascribed by a punter to each horse in a race will be conditional on the knowledge the punter has of the past form of each of the horses. What is more, those probabilities will be subject to change by the punter in the light of new evidence, when, for example, he finds on arrival at the racetrack that one of the horses is sweating badly and looking decidedly sick. Bayes' theorem is a theorem prescribing how probabilities are to be changed in the light of new evidence.

In the context of science the issue is how to ascribe probabilities to theories or hypotheses in the light of evidence. Let $P(\mathbf{h}/\mathbf{e})$ denote the probability of a hypothesis **h** in the light of evidence **e**, $P(\mathbf{e}/\mathbf{h})$ denote the probability to be ascribed to the evidence **e** on the assumption that the hypothesis **h** is correct, $P(\mathbf{h})$ the probability ascribed to **h** in the absence of knowledge of **e**, and $P(\mathbf{e})$ the probability ascribed to **e** in the absence of any assumption about the truth of **h**. Then Bayes' theorem can be written:

$$P(\mathbf{h}/\mathbf{e}) = P(\mathbf{h}).\frac{P(\mathbf{e}/\mathbf{h})}{P(\mathbf{e})}$$

$P(\mathbf{h})$ is referred to as the *prior probability*, since it is the probability ascribed to the hypothesis prior to consideration of the evidence, **e**, and $P(\mathbf{h}/\mathbf{e})$ is referred to as the *posterior probability*, the probability after the evidence, **e**, is taken into account. So the formula tells us how to change the probability of a hypothesis to some new, revised probability in the light of some specified evidence.

The formula indicates that the prior probability, P(**h**), is to be changed by a scaling factor P(**e/h**)/P(**e**) in the light of evidence **e**. It can readily be seen how this is in keeping with common intuitions. The factor P(**e/h**) is a measure of how likely **e** is given **h**. It will take a maximum value of 1 if **e** follows from **h** and a minimum value of zero if the negation of **e** follows from **h**. (Probabilities always take values in between 1, representing certainty, and zero, representing impossibility.) The extent to which some evidence supports a hypothesis is proportional to the degree to which the hypothesis predicts the evidence, which seems reasonable enough. The term in the divisor of the scaling factor, P(**e**), is a measure of how likely the evidence is considered to be when the truth of the hypothesis, **h**, is not assumed. So, if some piece of evidence is considered extremely likely whether we assume a hypothesis or not, the hypothesis is not supported significantly when that evidence is confirmed, whereas if that evidence is considered very unlikely unless the hypothesis is assumed, then the hypothesis will be highly confirmed if the evidence is confirmed. For instance, if some new theory of gravitation were to predict that heavy objects fall to the ground, it would not be significantly confirmed by the observation of the fall of a stone, since the stone would be expected to fall anyway. On the other hand, if that new theory were to predict some small variation of gravity with temperature, then the theory would be highly confirmed by the discovery of that effect, since it would be considered most unlikely in the absence of the new theory.

An important aspect of the Bayesian theory of science is that the calculations of prior and posterior probabilities always take place against a background of assumptions that are taken for granted, that is, assuming what Popper called background knowledge. So, for example, when it was suggested in the previous paragraph that P(**e/h**) takes the value 1 when **e** follows from **h**, it was taken for granted that **h** was to be taken in conjunction with the available background knowledge. We have seen in earlier chapters that theories need to be augmented by suitable auxiliary assumptions before they yield testable predictions. The Bayesians take these considerations on board. Throughout this discussion it is assumed that probabilities are calculated against a background of assumed knowledge.

It is important to clarify in what sense Bayes' theorem is indeed a *theorem*. Although we will not consider the details here, we note that there are some minimal assumptions about the nature of probability which taken together constitute the so-called probability calculus. These assumptions are accepted by Bayesians and non-Bayesians alike. It can be shown that denying them has a range of undesirable consequences. It can be shown, for example, that a gambling system that violates the probability calculus is 'irrational' in the sense that it makes it possible for wagers to be placed on all possible outcomes of a game, race or whatever in such a way that the participants on one or other side of the betting transaction will win *whatever the outcome.* (Systems of betting odds that allow this possibility are called Dutch Books. They violate the probability calculus.) Bayes' theorem can be derived from the premises that constitute the probability calculus. In that sense, the theorem in itself is uncontentious.

So far, we have introduced Bayes' theorem, and have tried to indicate that the way in which it prescribes that the probability of a hypothesis be changed in the light of evidence captures some straightforward intuitions about the bearing of evidence on theories. Now we must press the question of the interpretation of the probabilities involved more strongly.

Subjective Bayesianism

The Bayesians disagree among themselves on a fundamental question concerning the nature of the probabilities involved. On one side of the division we have the 'objective' Bayesians. According to them, the probabilities represent probabilities that rational agents *ought* to subscribe to in the light of the objective situation. Let me try to indicate the gist of their position with an example from horse racing. Suppose we are confronted by a list of the runners in a horse race and we are given no information about the horses at all. Then it might be argued that on the basis of some 'principle of indifference' the only rational way of ascribing probabilities to the likelihood of each horse winning is to distribute the probabilities equally among the runners. Once we have these 'objective' prior probabilities to start with, then Bayes' theorem dictates how the probabilities are to be modified in the light of any evidence, and so the posterior probabilities that result are also those that a rational agent *ought* to accept. A major, and notorious, problem with this approach, at least in the domain of science, concerns how to ascribe objective prior probabilities to hypotheses. What seems to be necessary is that we list all the possible hypotheses in some domain and distribute probabilities among them, perhaps ascribing the same probability to each employing the principle of indifference. But where is such a list to come from? It might well be thought that the number of possible hypotheses in any domain is infinite, which would yield zero for the probability of each and the Bayesian game cannot get started. All theories have zero probability and Popper wins the day. How is some finite list of hypotheses enabling some objective distribution of non-zero prior probabilities to be arrived at? My own view is that this problem is insuperable, and I also get the impression from the current literature that most Bayesians are themselves coming around to this point of view. So let us turn to 'subjective' Bayesianism.

For the subjective Bayesian the probabilities to be handled by Bayes' theorem represent subjective degrees of belief. They argue that a consistent interpretation of probability theory can be developed on this basis, and, moreover, that it is an interpretation that can do full justice to science. Part of their rationale can be grasped by reference to the examples I invoked in the opening paragraph of this chapter. Whatever the strength of the arguments for attributing zero probability to all hypotheses and theories, it is simply not the case, argue the subjective Bayesians, that people in general and scientists in particular ascribe zero probabilities to well-confirmed theories. The fact that I pre-booked my trip to the mountains to observe Halley's comet suggests that they are right in my case at least. In their work, scientists take many laws for granted. The unquestioning use of the law of refraction of light by astronomers and Newton's laws by those involved in the space program demonstrates that they ascribe to those laws a probability close, if not equal, to unity. The subjective Bayesians simply take the degrees of belief in hypotheses that scientists as a matter of fact happen to have as the basis for the prior probabilities in their Bayesian calculations. In this way they escape Popper's strictures to the effect that the probability of all universal hypotheses must be zero.

Bayesianism makes a great deal of sense in the context of gambling. We have noted that adherence to the probability calculus, within which Bayes' theorem can be proved, is a sufficient condition to avoid Dutch Books. Bayesian approaches to science capitalise on this by drawing a close analogy between science and gambling systems. The degree of belief held by a scientist in a hypothesis is analogous to the odds on a particular horse winning a race that he or she considers to be fair. Here there is a possible source of ambiguity that needs to be addressed. If we stick to our analogy with horse racing, then the odds considered to be fair by punters can be taken as referring either to their private subjective degrees of belief or to their beliefs as expressed in practice in their betting behaviour. These are not necessarily the same thing. Punters can depart from the dictates of the odds they believe in by becoming flustered at the race-track or by losing their nerve when the system of odds they believe in warrants a particularly large bet. Not all Bayesians make the same choice between these alternatives when applying the Bayesian calculus to science. For example, Jon Dorling (1979) takes the probabilities to measure what is reflected in scientific practice and Howson and Urbach (1993) take them to measure subjective degrees of belief. A difficulty with the former stand is knowing what it is within scientific practice that is meant to correspond to betting behaviour. Identifying the probabilities with subjective degrees of belief, as Howson and Urbach do, at least has the advantage of making it clear what the probabilities refer to.

Attempting to understand science and scientific reasoning in terms of the subjective beliefs of scientists would seem to be a disappointing departure for those who seek an objective account of science. Howson and Urbach have an answer to that charge. They insist that the Bayesian theory constitutes an *objective* theory of scientific inference. That is, given a set of prior probabilities and some new evidence, Bayes' theorem dictates in an objective way what the new, posterior, probabilities must be in the light of that evidence. There is no difference in this respect between Bayesianism and deductive logic, because logic has nothing to say about the source of the propositions that constitute the premises of a deduction either. It simply dictates what follows from those propositions once they are given. The Bayesian defence can be taken a stage further. It can be argued that the beliefs of individual scientists, however much they might differ at the outset, can be made to converge given the appropriate input of evidence. It is easy to see in an informal way how this can come about. Suppose two scientists start out by disagreeing greatly about the probable truth of hypothesis **h** which predicts otherwise unexpected experimental outcome **e**. The one who attributes a high probability to **h** will regards **e** as less unlikely than the one who attributes a low probability to **h**. So P(**e**) will be high for the former and low for the latter. Suppose now that **e** is experimentally confirmed. Each scientist will have to adjust the probabilities for **h** by the factor P(**e/h**)/P(**e**). However, since we are assuming that **e** follows from **h**, P(**e/h**) is 1 and the scaling factor is 1/P(**e**). Consequently, the scientist who started with a low probability for **h** will scale up that probability by a larger factor than the scientist who started with a high probability for **h**. As more positive evidence comes in, the original doubter is forced to scale up the probability in such a way that it eventually approaches that of the already convinced scientist. In this kind of way, argue the Bayesians, widely differing subjective opinions can be brought into conformity in response to evidence in an objective way.

Applications of the Bayesian formula

The preceding paragraph has given a strong foretaste of the kind of ways in which the Bayesians wish to capture and sanction typical modes of reasoning in science. In this section we will sample some more examples of Bayesianism in action.

In earlier chapters it was pointed out that there is a law of diminishing returns at work when testing a theory against experiment. Once a theory has been confirmed by an experiment once, repeating that same experiment under the same circumstances will not be taken by scientists as confirming the theory to as high a degree as the first experiment did. This is readily accounted for by the Bayesian. If the theory **T** predicts the experimental result

E then the probability P(**E/T**) is 1, so that the factor by which the probability of **T** is to be increased in the light of a positive result **E** is 1/P(**E**). Each time the experiment is successfully performed, the more likely the scientist will be to expect it to be performed successfully again the subsequent time. That is, P(**E**) will increase. Consequently, the probability of the theory being correct will increase by a smaller amount on each repetition.

Other points in favour of the Bayesian approach can be made in the light of historical examples. Indeed, I suggest that it is the engagement by the Bayesians with historical cases in science that has been a key reason for the rising fortunes of their approach in recent years, a trend begun by Jon Dorling (1979). In our discussion of Lakatos's methodology we noted that according to that methodology it is the confirmations of a program that are important rather than the apparent falsifications, which can be blamed on the assumptions in the protective belt rather than on the hard core. The Bayesians claim to be able to capture the rationale for this strategy. Let us see how they do it, by looking at a historical example utilised by Howson and Urbach (1993, pp. 136–42).

The example concerns a hypothesis put forward by William Prout in 1815. Prout, impressed by the fact that atomic weights of the chemical elements relative to the atomic weight of hydrogen are in general close to whole numbers, conjectured that atoms of the elements are made up of whole numbers of hydrogen atoms. That is, Prout saw hydrogen atoms as playing the role of elementary building blocks. The question at issue is what the rational response was for Prout and his followers to the finding that the atomic weight of chlorine relative to hydrogen (as measured in 1815) was 35.83, that is, not a whole number. The Bayesian strategy is to assign probabilities that reflect the prior probabilities that Prout and his followers might well have assigned to their theory together with relevant aspects of background knowledge, and then use Bayes' theorem to calculate how these probabilities change in light of the discovery of the problematic evidence, namely, the non-integral value of the atomic weight of chlorine. Howson and Urbach attempt to show that when this is done the result is that the probability of Prout's hypothesis falls just a little, whereas the probability of the relevant measurements being accurate falls dramatically. In light of this it seems quite reasonable for Prout to have retained his hypothesis (the hard core) and to have put the blame on some aspect of the measuring process (the protective belt). It would seem that a clear rationale has been given for what in Lakatos's methodology appeared as 'methodological decisions' that were not given any grounding. What is more, it would seem that Howson and Urbach, who are following the lead of Dorling here, have given a general solution to the so-called Duhem-Quine problem. Confronted with the problem of which part of a web of assumptions to blame for an apparent falsification, the Bayesian answer is to feed in the appropriate prior probabilities and calculate the posterior probabilities. These will show which assumptions slump to a low probability, and consequently which assumptions should be dropped to maximise the chances of future success.

I will not go through the details of the calculations in the Prout case, or any of the other examples that Bayesians have given, but I will say enough to at least give the flavour of the way in which they proceed. Prout's hypothesis, **h**, and the effect of the evidence, **e**, the non-integral atomic weight of chlorine, on the probability to be assigned to it is to be judged in the context of the available background knowledge, **a**. The most relevant aspect of the background knowledge is the confidence to be placed in the available techniques for measuring atomic weights and the degree of purity of the chemicals involved. Estimates need to be made about the prior probabilities of **h, a** and **e**. Howson and Urbach suggest a value of 0.9 for P(**h**), basing their estimate on historical evidence to the effect that the Proutians were very convinced of the truth of their hypothesis. They place P(**a**) somewhat lower at 0.6, on the grounds that chemists were aware of the problem of impurities, and that there were variations in the results of different measurements of the atomic weight of particular elements. The probability P(**e**) is assessed on the assumption that the alternative to **h** is a random distribution of atomic weights, so, for instance, P(**e/not h** & **a**) is ascribed a probability 0.01 on the grounds that, if the atomic weight of chlorine is randomly distributed over a unit interval it would have a one in a hundred chance of being 35.83, rather than 35.82, 35.61 or any of the hundred possibilities between 35.00 and 36.00. These probability estimates, and a few others like them, are fed into Bayes' theorem to yield posterior probabilities, P(**h/e**) and P(**a/e**), for **h** and **a**. The result is 0.878 for the former and 0.073 for the latter. Note that the probability for **h**, Prout's hypothesis, has fallen only a small amount from the original 0.9, whereas the probability of **a**, the assumption that the measurements are reliable, has fallen dramatically from 0.6 to 0.073. A reasonable response for the Proutians, conclude Howson and Urbach, was to retain their hypothesis and doubt the measurements. They point out that nothing much hinges on the absolute value of the numbers that are fed into the calculation so long as they are of the right kind of order to reflect the attitudes of the Proutians as reflected in the historical literature.

The Bayesian approach can be used to mount a criticism of some of the standard accounts of the undesirability of ad hoc hypotheses and related issues. Earlier in this book I proposed the idea, following Popper, that ad hoc hypotheses are undesirable because they are not testable independently of the evidence that led to their formulation. A related idea is that evidence that is used to construct a theory cannot be used again as evidence for that theory. From the Bayesian point of view, although these notions sometimes yield appropriate answers concerning how well theories are confirmed by evidence, they also go astray, and, what is more, the rationale underlying them is misconceived. The Bayesians attempt to do better in the following kinds of ways.

Bayesians agree with the widely held view that a theory is better confirmed by a variety of kinds of evidence than by evidence of a particular kind. There is a straightforward Bayesian rationale that explains why this should be so. The point is that there are diminishing returns from efforts to confirm a theory by a single kind of evidence. This follows from the fact that each time the theory is confirmed by that kind of evidence, then the probability expressing the degree of belief that it will do so in the future gradually increases. By contrast, the prior probability of a theory being confirmed by some new kind of evidence may be quite low. In such cases, feeding the results of such a confirmation, once it occurs, into the Bayesian formula leads to a significant increase in the probability ascribed to the theory. So the significance of independent evidence is not in dispute. Nevertheless, Howson and Urbach urge that, from the Bayesian point of view, if hypotheses are to be dismissed as ad hoc, the absence of independent testability is not the right reason for doing so. What is more, they deny that data used in the construction of a theory cannot be used to confirm it.

A major difficulty with the attempt to rule out ad hoc hypotheses by the demand for independent testability is that it is too weak, and admits hypotheses in a way that at least clashes with our intuitions. For instance, let us consider the attempt by Galileo's rival to retain his assumption that the moon is spherical in the face of Galileo's sightings of its moons and craters by proposing the existence of a transparent, crystalline substance enclosing the observable moon. This adjustment cannot be ruled out by the independent testability criterion because it was independently testable, as evidenced by the fact that it has been refuted by the lack of interference from any such crystalline spheres experienced during the various moon landings. Greg Bamford (1993) has raised this, and a range of other difficulties with a wide range of attempts to define the notion of ad hocness by philosophers in the Popperian tradition, and suggest that they are attempting to define a technical notion for what is in effect nothing more than a commonsense idea. Although Bamford's critique is not from a Bayesian point of view, the response of Howson and Urbach is similar, insofar as their view is that ad hoc hypotheses are rejected simply because they are considered implausible, and are credited with a low probability because of this. Suppose a theory **t** has run into trouble with some problematic evidence and is modified by adding assumption **a**, so that the new theory, **t**, is (**t** & **a**). Then it is a straightforward result of probability theory that P(**t** & **a**) cannot be greater than P(**a**). From a Bayesian point of view, then, the modified theory will be given a low probability simply on the grounds that P(**a**) is unlikely. The theory of Galileo's rival could be rejected to the extent that his suggestion was implausible. There is nothing more to it, and nothing else needed.

Let us now turn to the case of the use of data to construct a theory, and the denial that that data can be considered to support it. Howson and Urbach (1993, pp. 154–7) suggest counter examples. Consider an urn containing counters, and imagine that we begin with the assumption that all of the counters are white and none of them coloured. Suppose we now draw counters 1,000 times, replacing the counter and shaking the urn after each draw, and that the result is that 495 of the counters are white. We now adjust our hypothesis to be that the urn contains white and coloured counters in equal numbers. Is this adjusted hypothesis supported by the evidence used to arrive at the revised, equal numbers, hypothesis? Howson and Urbach suggest, reasonably, that it is, and show why this is so on Bayesian grounds. The crucial factor that leads to the probability of the equal numbers hypothesis increasing as a result of the experiment that drew 495 white counters is the probability of drawing that number if the equal numbers hypothesis is false. Once it is agreed that that probability is small, the result that the experiment confirms the equal numbers hypothesis follows straightforwardly from the Bayesian calculus, even though the data were used in the construction of the hypothesis.

There is a standard criticism often levelled at the Bayesian approach that does strike at some versions of it, but I think the version defended by Howson and Urbach can counter it. To utilise Bayes' theorem it is necessary to be able to evaluate P(**e**), the prior probability of some evidence that is being considered. In a context where hypothesis **h** is being considered, it is convenient to write P(**e**) as P(**e/h**).P(**h**) + P(**e/not h**).P(**not h**), a straightforward identity in probability theory. The Bayesian needs to be able to estimate the probability of the evidence assuming the hypothesis is true, which may well be unity if the evidence follows from the hypothesis, but also the probability of the evidence should the hypothesis be false. It is this latter factor that is the problematic one. It would appear that it is necessary to estimate the likelihood of the evidence in the light of all hypotheses other than **h**. This is seen as a major obstacle, because no particular scientist can be in a position to know all possible alternatives to **h**, especially if, as some have suggested, this must include all hypotheses not yet invented. The response open to Howson and Urbach is to insist that the probabilities in their Bayesian calculus represent personal probabilities, that is, the probabilities that individuals, as a matter of fact, attribute to various propositions. The value of the probability of some evidence being true in the light of alternatives to **h** will be decided on by a scientist in the light of what that scientist happens to know (which will certainly exclude hypotheses not yet invented). So, for instance, when dealing with the Prout case, Howson and Urbach take the only alternative to Prout's hypothesis to be the hypothesis that atomic weights are randomly distributed on the basis of historical evidence to the effect that that is what the Proutians believed to be the alternative. It is the thoroughgoing nature of their move to subjective probabilities that makes it possible for Howson and Urbach to avoid the particular problem raised here.

In my portrayal of the elements of the Bayesian analysis of science, I have concentrated mainly on the position outlined by Howson and Urbach because it seems to me to be the one most free of inconsistencies. Because of the way in which probabilities are interpreted in terms of degrees of the beliefs actually held by scientists, their system enables non-zero probabilities to be attributed to theories and hypotheses, it gives a precise account of how the probabilities are to be modified in the light of evidence, and it is able to give a rationale for what many take to be key features of scientific method. Howson and Urbach embellish their system with historical case studies.

Critique of subjective Bayesianism

As we have seen, subjective Bayesianism, the view that consistently understands probabilities as the degrees of belief actually held by scientists, has the advantage that it is able to avoid many of the problems that beset alternative Bayesian accounts that seek for objective probabilities of some kind. For many, to embrace subjective probabilities is to pay too high a price for the luxury of being able to attribute probabilities to theories. Once we take probabilities as subjective degrees of belief to the extent that Howson and Urbach, for example, urge that we do, then a range of unfortunate consequences follow.

The Bayesian calculus is portrayed as an objective mode of inference that serves to transform prior probabilities into posterior probabilities in the light of given evidence. Once we see things in this way, it follows that any disagreements in science, between proponents of rival research programs, paradigms or whatever, reflected in the (posterior) beliefs of scientists, must have their source in the prior probabilities held by the scientists, since the evidence is taken as given and the inference considered to be objective. But the prior probabilities are themselves totally subjective and not subject to a critical analysis. They simply reflect the various degrees of belief each individual scientist happens to have. Consequently, those of us who raise questions about the relative merits of competing theories and about the sense in which science can be said to progress will not have our questions answered by the subjective Bayesian, unless we are satisfied with an answer that refers to the beliefs that individual scientists just happen to have started out with.

If subjective Bayesianism is the key to understanding science and its history, then one of the most important sources of information that we need to have access to in order to acquire that understanding is the degrees of belief that scientists actually do or did hold. (The other source of information is the evidence, which is discussed below.) So, for instance, an understanding of the superiority of the wave theory over the particle theory of light will require some knowledge of the degrees of belief that Fresnel and Poisson, for instance, brought to the debate in the early 1830s. There are two problems here. One is the problem of gaining access to a knowledge of these private degrees of belief. (Recall that Howson and Urbach distinguish between private beliefs and actions and insist that it is the former with which their theory deals, so we cannot infer beliefs of scientists from what they do, or even write.) The second problem is the implausibility of the idea that we need to gain access to these private beliefs in order to grasp the sense in which, say, the wave theory of light was an improvement on its predecessor. The problem is intensified when we focus on the degree of complexity of modern science, and the extent to which it involves collaborative work. (Recall my comparison with workers constructing a cathedral in chapter 8.) An extreme, and telling, example is provided by Peter Galison's (1997) account of the nature of the work in current fundamental particle physics, where very abstruse mathematical theories are brought to bear on the world via experimental work that involves elaborate computer techniques and instrumentation and requires state-of-the-art engineering for its operation. In situations like this there is no single person who grasps all aspects of this complex work. The theoretical physicist, the computer programmer, the mechanical engineer and the experimental physicist all have their separate skills, which are brought to bear on a collaborative enterprise. If the progressiveness of this enterprise is to be understood as focusing on degrees of belief, then whose degree of belief do we choose and why?

The extent to which degrees of belief are dependent on prior probabilities in Howson and Urbach's analysis is the source of another problem. It would seem that, provided a scientist believes strongly enough in his or her theory to begin with (and there is nothing in subjective Bayesianism to prevent degrees of belief as strong as one might wish), then this belief cannot be shaken by any evidence to the contrary, however strong or extensive it might be. This point is in fact illustrated by the Prout study, the very study that Howson and Urbach use to support their position. Recall that in that study we assume that the Proutians began with a prior probability of 0.9 for their theory that atomic weights are equal multiples of the atomic weight of hydrogen and a prior probability of 0.6 for the assumption that atomic weight measurements are reasonably accurate reflections of actual atomic weights. The posterior probabilities, calculated in the light of the 35.83 value obtained for chlorine, were 0.878 for Prout's theory and 0.073 for the assumption that the experiments are reliable. So the Proutians were right to stick to their theory and reject the evidence. I point out here that the original incentive behind Prout's hypothesis was the near integral values of a range of atomic weights other than chlorine, measured by the very techniques that the Proutians have come to regard as so unreliable that they warrant a probability as low as 0.073! Does this not show that if scientists are dogmatic enough to begin with they can offset any adverse evidence? Insofar as it does, there is no way that the subjective Bayesian can identify such activity as bad scientific practice. The prior probabilities cannot be judged. They must be taken as simply given. As Howson and Urbach (1993, p. 418) observe, 'how prior distributions are determined is simply not something that comes within the scope of [their] theory'.

Bayesians seem to have a counter to the Popperian claim that the probability of all theories must be zero, insofar as they identify probabilities with the degrees of belief that scientists happen, as a matter of fact, to possess. However, the Bayesian position is not that simple. For it is necessary for the Bayesians to ascribe probabilities that are *counterfactual*, and so cannot be simply identified with degrees of belief actually held. Let us take the problem of how past evidence is to count for a theory as an example. How can the observations of Mercury's orbit be taken as confirmation of Einstein's theory of general relativity, given that the observations preceded the theory by a number of decades? To calculate the probability of Einstein's theory in the light of this evidence, the subjective Bayesian is required, among other things, to provide a measure for the probability an Einstein supporter would have given to the probability of Mercury's orbit precessing in the way that it does *without a knowledge of Einstein's theory*. That probability is not a measure of the degree of belief that a scientist actually has but a measure of a degree of belief they would have had if they did not know what they in fact do know. The status of these degrees of belief, and the problem of how one is to evaluate them, pose serious problems, to put it mildly.

Let us now turn to the nature of 'evidence' as it figures in subjective Bayesianism. We have treated the evidence as a given, something that is fed into Bayes' theorem to convert prior probabilities to posterior probabilities. However, as the discussion of the early chapters of this book should have made clear, evidence in science is far from being straightforwardly given. The stand taken by Howson and Urbach (1993, pp. 406–7) is explicit and totally in keeping with their overall approach.

> The Bayesian theory we are proposing is a theory of inference from data; we say nothing about whether it is correct to accept the data or even whether your commitment to the data is absolute. It may not be, and you may be foolish to repose in it the confidence you actually do. The Bayesian theory of support is a theory of how the acceptance as true of some evidential statement affects your belief in some hypothesis. How you come to accept the truth of the evidence and whether you are correct in accepting it as

true are matters which, from the point of view of the theory, are simply irrelevant.

Surely this is a totally unacceptable position for those who purport to be writing a book on *scientific reasoning*. For is it not the case that we seek an account of what counts as appropriate evidence in science? Certainly a scientist will respond to some evidential claim, not by asking the scientist making the claim how strongly he or she believes it, but by seeking information on the nature of the experiment that yielded the evidence, what precautions were taken, how errors were estimated and so on. A good theory of scientific method will surely be required to give an account of the circumstances under which evidence can be regarded as adequate, and be in a position to pinpoint standards that empirical work in science *should* live up to. Certainly experimental scientists have plenty of ways of rejecting shoddy work, and not by appealing to subjective degrees of belief.

Especially when they are responding to criticism, Howson and Urbach stress the extent to which both the prior probabilities and the evidence which need to be fed into Bayes' theorem are subjective degrees of belief about which the subjective Bayesian has nothing to say. But to what extent can what remains of their position be called a theory of scientific method? All that remains is a theorem of the probability calculus. Suppose we concede to Howson and Urbach that this theorem, as interpreted by them, is indeed a theorem with a status akin to deductive logic. Then this generous concession serves to bring out the limitation of their position. Their theory of scientific method tells us as much about science as the observation that science adheres to the dictates of deductive logic. The vast majority, at least, of philosophers of science would have no problem accepting that science takes deductive logic for granted, but would wish to be told much more.

Further reading

Dorling (1979) was an influential paper that put subjective Bayesianism on its modern trend, and Howson and Urbach (1993) is a sustained and unabashed case for it. Horwich (1982) is another attempt to understand science in terms of subjective probability. Rosenkrantz (1977) is an attempt to develop a Bayesian account of science involving objective probabilities. Earman (1992) is a critical, but technical, defence of the Bayesian program. Mayo (1996) contains a sustained critique of Bayesianism.

CHAPTER 13

The new experimentalism

Introduction

If we regard the Bayesian account of scientific inference as a failure, we still have not provided much by way of some characterisation of what it is that is distinctive about scientific knowledge. Popper posed problems for positivism and inductivism by stressing the theory-dependence of observation and the extent to which theories always transcend, and so can never be derived from, the evidence. Popper's account of science was based on the idea that the best theories are those that survive the severest tests. However, his account was unable to give clear guidance as to when a theory, rather than some element of background knowledge, should be held responsible for a failed test, and was unable to say something sufficiently positive about theories that happen to have survived tests. The subsequent attempts that we discussed all involved taking the idea of theory-dependence *further* than Popper did. Lakatos introduced research programs, and saw them retained or rejected according to conventional decisions, decisions, for example, to blame auxiliary assumptions rather than hard-core principles for apparent falsifications. However, he was unable to give grounds for those decisions, and in any case they were too weak to specify when it was time to abandon a research program in favour of another. Kuhn introduced paradigms rather than research programs, thus introducing a degree of paradigm-dependence in science that was more far-reaching than Popper's theory-dependence, so much so that Kuhn was even worse off than Lakatos in giving a clear answer to the question of the sense in which a paradigm could be said to be an improvement on the one it replaced. Feyerabend can be seen as taking the theory-dependence movement to its extreme, giving up on the idea of special methods and standards for science altogether, and joining Kuhn in the portrayal of rival theories as incommensurable. The Bayesians can also be seen as part of what I am calling the theory-dependence tradition. For them the background theoretical assumptions that inform the judgments about the merits of scientific theories are brought in by way of the prior probabilities.

For one group of philosophers, the range of problems that beset contemporary philosophy of science are to be confronted by tackling the move towards radical theory-dependence at its source. Although they do not wish to return to the positivist idea that the senses provide an unproblematic basis for science, they do seek a relatively secure basis for science, not in observation but in experiment. I shall follow Robert Ackermann (1989) and refer to this recent trend as 'the new experimentalism'. According to its proponents, experiment can, in the words of Ian Hacking (1983, p. vii) have a 'life of its own' independent of large-scale theory. It is argued that experimentalists have a range of practical strategies for establishing the reality of experimental effects without needing recourse to large-scale theory. What is more, if scientific progress is seen as the steady build-up of the stock of experimental knowledge, then the idea of cumulative progress in science can be reinstated and is not threatened by claims to the effect that there are scientific revolutions involving large-scale theory change.

Experiment with life of its own

We begin this section with a historical story, drawing heavily on Gooding (1990). Late in the summer of 1820 reports reached Britain of Oersted's finding that the magnetic effect of a current-carrying wire in some way circulates the wire. Faraday undertook experimental work to clarify what this claim amounted to and to develop it further. Within a few months he had constructed what was, in effect, a primitive electric motor. A cylindrical glass tube was sealed by corks, top and bottom. A wire running through the centre of the top cork into the cylinder ended in a hook from which a second wire hung vertically. Its lower end was free to rotate around the tip of a soft iron cylinder protruding into the base of the cylinder via the bottom cork. Electrical contact between the lower tip of the dangling wire and the iron core was maintained via a pool of mercury resting on the lower cork. To activate this 'motor', one pole of a bar magnet was held adjacent to the end of the iron core emerging from the lower cork, while a conducting wire connected the iron core to the wire emerging from the top cork via an electric cell. The ensuing current caused the lower tip of the dangling wire to rotate around the magnetised iron core, maintaining contact with the mercury as it did so. Faraday promptly sent a sample of this device to his rivals around Europe, complete with instructions on how to make it work. He pointed out to them that they could reverse the direction of the rotation either by reversing the connections to the battery or by reversing the magnet.

Is it useful or appropriate to regard this accomplishment of Faraday's as theory-dependent and fallible? It can be said to be theory-dependent in a very weak sense. Faraday's rivals on the Continent would not have been able to follow his instructions if they did not know what a magnet, mercury and an electric cell were. But this amounts to no more than a refutation of the extreme empiricist idea that facts must be established directly by the entry of sensory data into a mind that otherwise knows nothing. Nobody need deny the claim that someone who cannot tell the difference between a magnet and a carrot is not in a position to appreciate what counts as an established fact in electromagnetism. It is surely injudicious to use the term 'theory' in such a general sense that 'carrots are not magnets' becomes a theory. What is more, construing all talk as 'theory-dependent' does not help get to grips with the genuine differences between the likes of Faraday and Ampere. Faraday, as is well known, sought to understand electric and magnetic phenomena in terms of lines of force emanating from electrically charged bodies and magnets and filling the space around them, while theorists on the Continent thought of electric fluids residing in insulators and flowing through conductors, with elements of fluid acting on each other at a distance. These were the theories at stake, and the appreciation of Faraday's motor effect was not 'theory-dependent' in the sense that an appreciation of it depended on the acceptance of or familiarity with some version of one of the rival theories. Within electromagnetism at the time

Faraday's motor constituted an experimentally established theory-neutral effect which all electromagnetic theories were obliged to take account of.

Nor is it helpful to regard Faraday's motor effect as fallible. It is true that Faraday's motors sometimes do not work, because the magnet is too weak or because the wire is immersed so far into the mercury that the latter offers too much resistance to rotation, or whatever. Consequently, the statement 'all wires situated in an experimental arrangement meeting Faraday's description rotate' is false. But this simply indicates that attempting to capture the essence of Faraday's discovery with universal statements of this kind is inappropriate. Faraday discovered a new experimental effect, demonstrated it by constructing a version of his device that did work, and gave instructions to his rivals that enabled them to build devices that worked too. The odd failure is neither surprising nor relevant. The theoretical explanation of Faraday's motor that would be accepted today differs from that offered by both Faraday and Ampere in significant respects. But it remains the case that Faraday's motors usually work. It is difficult to comprehend how future advances in theory could somehow lead to the conclusion that electric motors don't work (although they might well be rendered obsolete by future discoveries of yet other experimental effects). Looked at in this way, experimental effects that can be produced in a controlled way are not fallible, they are here for keeps. What is more, if we understand progress in science in terms of the accumulation of such effects, then we have a theory-independent understanding of its growth.

A second example supports further this way of looking at things. Jed Buchwald's (1989) detailed study of the experimental career of Heinrich Hertz indicates the extent to which Hertz aimed to produce novel experimental effects. Some of his claims to have done so did not meet with general acceptance. It is not difficult to appreciate why. Hertz had learnt his electromagnetism through Helmholtz and saw things in terms of Helmholtz's theoretical framework, which was just one of the several theoretical approaches to electromagnetism at the time (the chief alternatives being those of Weber and Maxwell). That the experimental findings of Hertz constituted novel effects could only be appreciated and defended if the fine details of the theoretical interpretation Hertz brought to his experiments were appreciated and defended. These results were highly theory-dependent, and this, a new experimentalist might well argue, is precisely why they were not generally accepted as constituting novel effects. Things were quite otherwise once Hertz had produced his electric waves. That there were such waves could be demonstrated in a way that was independent of which general theory was subscribed to. Hertz was able to exhibit this new effect in a controlled way. He set up standing waves and showed that small spark detectors showed maximum sparking at the antinodes and no sparking at the nodes of these waves. This was by no means easily achieved, nor were the results easily reproduced, as Buchwald found when he tried it. But I am not claiming the experiments were easy. I am simply claiming that the fact that the experiments demonstrated the existence of a new experimentally produced phenomenon could be appreciated in a way that did not rely on recourse to one or other of the competing electromagnetic theories, a claim borne out by the rapidity with which Hertz's waves were accepted by all camps.

The production of controlled experimental effects can be accomplished and appreciated independently of high-level theory, then. In a similar vein, the new experimentalist can point to a range of strategies available to experimenters for establishing their claims that do not involve appeal to high-level theory. Let us consider, for example, how an experimentalist might argue that a particular observation by way of an instrument represents something real rather than an artifact. Ian Hacking's (1983, pp. 186–209) stories concerning the use of microscopes illustrate the point well. A miniature grid, with labelled squares is etched on a piece of glass which is then photographically reduced to such an extent that the grid becomes invisible. The reduced grid is viewed through a microscope that reveals the grid, complete with labelled squares. This already is a strong indication that the microscope magnifies, and magnifies reliably – an argument, incidentally, that does not rely on a theory of how the microscope works. We now reflect on a biologist who is using an electron microscope to view red blood platelets mounted on our grid. (Here Hacking is reporting an actual sequence of affairs reported to him by a scientist.) Some dense bodies are observable within the cell. The scientist wonders if the bodies are present in the blood or are artifacts of the instrument. (He suspects the latter.) He notes which of the labelled squares on the grid contain these dense bodies. Next he views his sample through a fluorescence microscope. The same bodies appear once again, in the same locations on the grid. Can there be any doubt that what is being observed represents bodies in the blood rather than artifacts? All that is required to render this argument persuasive is the knowledge that the two microscopes work on quite different physical principles, so that the chance of both of them producing identical artifacts can be recognised as highly improbable. The argument does not require detailed theoretical knowledge of the workings of either instrument.

Deborah Mayo on severe experimental testing

Deborah Mayo (1996) is a philosopher of science who has attempted to capture the implications of the new experimentalism in a philosophically rigorous way. Mayo focuses on the detailed way in which claims are validated by experiment, and is concerned with identifying just what claims are borne out and how. A key idea underlying her treatment is that a claim can only be said to be supported by experiment if the various ways in which the claim could be at fault have been investigated and eliminated. A claim can only be said to be borne out by experiment if it has been severely tested by experiment, and a severe test of a claim, as usefully construed by Mayo, must be such that the claim would be unlikely to pass it if it were false.

Her idea can be illustrated by some simple examples. Suppose Snell's law of refraction of light is tested by some very rough experiments in which very large margins of error are attributed to the measurements of angles of incidence and refraction, and suppose that the results are shown to be compatible with the law within those margins of error. Has the law been supported by experiments that have severely tested it? From Mayo's perspective the answer is 'no' because, owing to the roughness of the measurements, the law of refraction would be quite likely to pass this test even if it were false and some other law differing not too much from Snell's law true. An exercise I carried out in my schoolteaching days serves to drive this point home. My students had conducted some not very careful experiments to test Snell's law. I then presented them with some alternative laws of refraction that had been suggested in antiquity and medieval times, prior to the discovery of Snell's law, and invited the students to test them with the measurements they had used to test Snell's law. Because of the wide margins of error they had attributed to their measurements, all of these alternative laws passed the test. This clearly brings out the point that the experiments in question did not constitute a severe test of Snell's law. That law would have passed the test even if it were false and one of the historical alternatives true.

A second example further illustrates the rationale behind Mayo's position. I had two cups of coffee this morning and this afternoon I have a headache. Is the claim 'my morning coffee caused me to have a headache' thereby confirmed? Mayo's position captures the reason why the answer is 'no'. Before the claim can be said to have been severely tested, and so confirmed, we must eliminate the various ways in which the claim could be in error. Perhaps my headache is due to the particularly strong Vietnamese beer I drank last night, to the fact that I got up too early, that I am finding this section particularly difficult to write, and so on. If some causal connection between coffee drinking and headaches is to be established then it will be necessary to conduct controlled experiments that will serve to eliminate other possible causes. We must seek to establish results that would be most unlikely to occur unless coffee does indeed cause headaches. An experiment constitutes support for a claim only if possible sources of error have been eliminated, so that the claim would be unlikely to pass the test unless it were true. This simple idea serves to capture some common intuitions about experimental reasoning in a neat way and is also extended by Mayo to offer some fresh insights.

Let us consider the so-called tacking paradox, which I illustrate with an example. Let us imagine Newton's theory, T, has been confirmed by carefully observing the motion of a comet, with care being taken to eliminate sources of error due to attraction from nearby planets, refraction in the earth's atmosphere and so on. Suppose that we now construct theory T' by tacking a statement such as 'emeralds are green' onto Newton's theory. Is T' confirmed by the observations of the comet? If we hold the view that a prediction, p, confirms a theory if p follows from the theory and is confirmed by experiment, then T' (and a vast number of similarly constructed theories) is confirmed by the observations in question, counter to our intuitions. Hence the 'tacking paradox'. However, T' is not confirmed from Mayo's point of view and the 'paradox' is dissolved. Given our assumptions about the elimination of possible sources of error, we can say that the orbit of the comet would be unlikely to have conformed to the Newtonian prediction unless Newton's theory were true. The same cannot be said about T' because the likelihood of the comet conforming to the Newtonian prediction would be totally unaltered if some emeralds were blue and hence T' false. T' is not confirmed by the experiment in question because that experiment does not probe the various ways in which 'emeralds are green' might be false. Observations of comets can severely test T but not T'.

Mayo extends this line of reasoning to less trivial cases. She is keen to keep theoretical speculation in check by identifying theoretical conclusions that go further beyond the experimental evidence than is warranted. Her analysis of Eddington's test of Einstein's prediction of the bending of light in a gravitational field illustrates the point.

Eddington took advantage of an eclipse of the sun to observe the relative position of stars in a situation where the light from them passed close to the sun on their passage to earth. He compared these relative positions with those observed later in the year, when the stars were no longer closely aligned with the sun. A measurable difference was detected. By looking at the details of the eclipse experiments Mayo is able to argue that Einstein's law of gravity, which is a consequence of his general theory of relativity, was confirmed by them, but the general theory of relativity itself was not. Let us see how she does so.

If the results of the eclipse experiments are to be taken as confirming the general theory of relativity, then it must be possible to argue that those results would be most unlikely to occur if the general theory is false. We must be able to eliminate erroneous links between the general theory and the results. This could not be done in the case in question because there are, as a matter of fact, a whole class of theories of space-time of which Einstein's theory is only one, all of which predict Einstein's law of gravity and hence the results of the eclipse experiments. If one of this class of theories other than Einstein's were true, and Einstein's false, exactly the same results of the eclipse experiments would be expected. Consequently, those experiments did not constitute a severe test of Einstein's general theory. They did not serve to distinguish between it and known alternatives. To claim that the eclipse experiments supported Einstein's general theory of relativity is to go beyond the experimental evidence further than is warranted.

The situation is different when we consider the more restricted claim that the eclipse experiments confirmed Einstein's law of gravity. The observations certainly were in conformity with that law, but before it is legitimate to take this as evidence for the law, we must eliminate other possible causes of the conformity. It is only then we can say that the observed displacements would not have occurred unless Einstein's law is true. Mayo shows in some detail how alternatives to Einstein's law, including Newtonian alternatives arising from an inverse square law of attraction between the sun and photons presumed to have mass, were considered and eliminated. Einstein's law of gravity was severely tested by the eclipse experiments in a way that the general theory of relativity was not.

The new experimentalists are generally concerned to capture a domain of experimental knowledge that can be reliably established independent of high-level theory. Mayo's position meshes well with that aspiration. From her perspective, experimental laws can be confirmed by severely testing them along the lines discussed above. The growth of scientific knowledge is to be understood as the accumulation and extension of such laws.

Learning from error and triggering revolutions

Experimental results confirm a claim when they can be argued to be free from error, and when the results would be unlikely if the claim were false. However, there is more to Mayo's focus on the importance of experimental error than this. She is concerned with how well-conducted experiments enable us to learn from error. Looked at from this point of view, an experiment that serves to detect an error in some previously accepted assertion serves a positive as well as a negative function. That is, it not only serves as a falsification of the assertion, but also positively identifies an effect not previously known. The positive role of error detection in science is well illustrated by Mayo's reformulation of Kuhn's notion of normal science.

Let us recall our account, in chapter 8, of the conflicting answers given by Popper and Kuhn to the question of why astrology fails to qualify as a science. According to Popper, astrology is not a science because it is unfalsifiable. Kuhn points out that this is inadequate because astrology was (and is) falsifiable. In the sixteenth and seventeenth centuries, when astrology was 'respectable', astrologers did make testable predictions, many of which turned out to be false. Scientific theories make predictions that turn out to be false too. The difference, according to Kuhn, is that science is in a position to learn constructively from the 'falsifications', whereas astrology was not. For Kuhn, there exists in normal science a puzzle-solving tradition that astrology lacked. There is more to science than the falsification of theories. There is also the way in which falsifications are constructively overcome. It is ironic, from this point of view, that Popper, who at times characterised his own approach with the slogan 'we learn from our mistakes', failed precisely because his negative, falsificationist account did not capture an adequate, positive account of how science learns from mistakes (falsifications).

Mayo sides with Kuhn here, identifying normal science with experimentation. Let us note some examples of the positive role played by error detection. The observation of the problematic features of Uranus's orbit posed problems for Newtonian theory in conjunction with the background knowledge of the time. But the positive side of the problem was the extent to which the source of the trouble could be traced, leading to the discovery of Neptune in the way we have already described. Another episode we have mentioned before concerns Hertz's experiments on cathode rays, which led him to conclude that they are not deflected by an electric field. J. J. Thomson was able to show that he was in error, in part by appreciating the extent to which the rays ionise the residual gas in discharge tubes, leading to a build-up of charged ions on electrodes and the formation of electric fields. By achieving lower pressures in his tubes and arranging his electrodes more appropriately, Thomson detected the influence of electric fields on cathode rays that Hertz had missed. But he had also learnt something about new effects concerning ionisation and the build-up of space charge. In the context of the deflection experiments these constituted impediments to be removed. However, they also turned out to be important in their own right. The ionisation of gases by the passage of charged particles through them was to be fundamental for the study of charged particles in cloud chambers. The experimentalist's detailed knowledge of the effects at work in an apparatus puts him or her in a position to be able to learn from error.

Mayo does more than simply translate Kuhn's notion of normal science into experimental practice. She points to the way in which the facility of experiment to detect and accommodate error can prove sufficient to trigger or contribute to a scientific revolution, a decidedly unKuhnian thesis. Mayo's best example concerns the experiments on Brownian Motion conducted by Jean Perrin towards the end of the first decade of last century. Perrin's detailed, ingenious, down-to-earth observations of the motions of Brownian particles established beyond reasonable doubt that their motion was random. This, together with observations of the variation of the density of the distribution of particles with height, enabled Perrin to show as conclusively as one could wish that the motion of the particles violate the second law of thermodynamics as well as conforming to detailed predictions of the kinetic theory. You can't get much more revolutionary than that. A similar story could be told about the way in which experimental investigations of black body radiation, radioactive decay and the photoelectric effect, for instance, forced an abandonment of classical physics and constituted important elements of the new quantum theory in the early decades of the twentieth century.

Implicit in the new experimentalist's approach is the denial that experimental results are invariably 'theory' or 'paradigm' dependent to the extent that they cannot be appealed to to adjudicate between theories. The reasonableness of this stems from the focus on experimental practice, on how instruments are used, errors eliminated, cross-checks devised and specimens manipulated. It is the extent to which this experimental life is sustained in a way that is independent of speculative theory that enables the products of that life to act as major constraints on theory. Scientific revolutions can be 'rational' to the extent that they are forced on us by experimental results. The extremes of the theory- or paradigm-dominated views of science have lost touch with, and cannot make sense of, one of its most distinctive components, experimentation.

The new experimentalism in perspective

The new experimentalists have shown how experimental results can be substantiated and experimental effects produced by an array of strategies involving practical interventions, cross-checking and error control and elimination in a way that can be, and typically is, independent of high-level theory. As a consequence of this, they are able to give an account of progress in science that construes it as the accumulation of experimental knowledge. Adopting the idea that the best theories are those that survive the severest tests, and understanding a severe experimental test of a claim as one that the claim is likely to fail if it is false, the new experimentalists can show how experiment can bear on the comparison of radically different theories, and also how experiment can serve to trigger scientific revolutions. Careful attention to the details of experiments and to exactly what they do establish serves to keep theorising in check, and helps to distinguish between what has been substantiated by experiment and what is speculative.

There is no doubt that the new experimentalism has brought philosophy of science down to earth in a valuable way, and that it stands as a useful corrective to some of the excesses of the theory-dominated approach. However, I suggest it would be a mistake to regard it as the complete answer to our question about the character of science. Experiment is not so independent of theory as the emphasis of the previous sections of this chapter might suggest.

The healthy and informative focus on the life of experiment should not blind us to the fact that theory has an important life too.

The new experimentalists are right to insist that to see every experiment as an attempt to answer a question posed by theory is a mistake that underestimates the extent to which experiment can have a life of its own. Galileo didn't have a theory about Jupiter's moons to test when he turned his telescope skywards, and, ever since then, many novel phenomena have been discovered by exploiting the opportunities opened up by new instruments or technologies. On the other hand, it does remain the case that theory often does guide experimental work and has pointed the way towards the discovery of novel phenomena. After all, it was a prediction of Einstein's theory of general relativity that motivated Eddington's eclipse expeditions and it was implications of the kinetic theory of gases that led Perrin to investigate Brownian motion in the way that he did. In a similar vein, it was fundamental theoretical issues concerned with whether the rate of change of the polarisation of dielectric media should have magnetic effects like a conduction current that put Hertz onto the experimental path that culminated in the production of radio waves, and Arago's discovery of the bright spot at the centre of a disc's shadow resulted from a direct test of Fresnel's wave theory of light.

Whether theory can sometimes guide the experimentalist in the right direction or not, the new experimentalists are keen to capture a sense in which experimental knowledge can be vindicated in a way that is independent of high-level theory. Certainly Deborah Mayo has given a detailed and convincing account of how experimental results can be reliably established using an array of error-eliminating techniques and error statistics. However, as soon as the need arises to attach significance to experimental results that extends beyond the experimental situations in which they were produced, then reference to theory needs to be made.

Mayo endeavours to show how error statistics can be applied to carefully controlled experiments to yield the conclusion that experiments of that type yield specified results with a (specified) high degree of probability. Recorded experimental results are treated as a sample of all the possible results that might be achieved by experiments of that type, and error statistics can be applied to attribute probabilities to the population on the basis of the sample. A basic issue here is the question of what counts as an experiment of the same type. All experiments will differ from one another in some respects, insofar as, for example, they are conducted at different times, in different laboratories, using different instruments and so on. The general answer to the query is that the experiments must be similar in relevant respects. However, judgments about what counts as relevant are made by drawing on current knowledge, and so are subject to change when that knowledge is improved. Imagine, for example, Galileo conducting a range of experiments from the results of which he concludes that acceleration due to gravity is a constant (and let us, counterfactually, allow Galileo the use of modern error statistics and imagine he is able to attribute a low probability to the possibility of a future experiment telling against him). From a modern standpoint one can see how Galileo's reliance on his value for the acceleration could let him down if, on some future occasion, he were working well above sea level. Working in a context where it was assumed that the tendency to fall is an inherent property of heavy objects that they possess just in virtue of their being material objects, as Galileo did, it is not apparent that height above sea level is relevant, and so not apparent that Galileo's sample is unrepresentative. Judgments about what counts as experiments of a similar type are made against a theoretical background.

Leaving that kind of problem aside, as soon as experimental results are considered to have a significance that goes beyond the specific conditions in which they were produced, theoretical considerations become crucial. This is evident, for example, in the way Deborah Mayo herself argues that the eclipse experiments confirmed Einstein's law of gravity. As Mayo explains, this involved showing that the results were incompatible with the best Newtonian estimates of the phenomena as well as any other alternative that could be dreamt up, such as Oliver Lodge's appeal to an ether mechanism. One by one, these alternatives were found wanting. Mayo (1996, p. 291) quotes with approval Dyson and Crommelin writing in an article in *Nature*: 'Hence we seem to be driven by exhaustion to the Einstein law as the only satisfactory explanation.' I do not wish to dispute that this shows that it was reasonable to accept Einstein's theory of gravity at the time in the light of these circumstances. But a crucial part of the argument rests on the assumption that there are, as a matter of fact, no acceptable alternatives. Mayo cannot rule out the possibility that there is some modification of Newtonian theory or an ether theory, not yet thought of, that is able to explain the results of the eclipse experiments. That is why she is wise not to attempt to attribute probabilities to hypotheses. So her argument for scientific laws and theories boils down to the claim that they have withstood severe tests better than any available competitor. The only difference between Mayo and the Popperians is that she has a superior version of what counts as a severe test. Theoretical considerations play a crucial role.

The new experimentalists insist that experimenters have at their disposal powerful techniques for establishing experimental knowledge in a robust and reliable way that can be relatively independent of rarefied theory. To the extent that these claims can be secured, it would seem that the excesses of fallibilism can be curbed and a cumulative account of scientific progress, understood as the growth of reliable experimental knowledge, defended. However, once theoretical considerations of the kind I have discussed in this section are admitted to play a crucial role, then a corresponding degree of fallibilism must be admitted.

The new experimentalism has not shown how theory, sometimes high-level theory, can be eliminated from science. It is significant, in this connection, to note that an important factor in decisions about the reliability of Newtonian mechanics in the context of space flights is the extent to which, given the expected speeds, deviations from it can be shown to be negligible *in the light of relativity theory.* It is undoubtedly the case that there is an important 'life of theory' in science. The principles of quantum mechanics, employed, for instance, in the refinements of the electron microscope or even the conservation of energy, used throughout science, are much more than generalisations from specified experiments. What kind of life do they have in science, and how is that life related to experiment?

Some of the new experimentalists seem to wish to draw a line between well-established experimental knowledge on the one hand and high-level theory on the other. (Deborah Mayo seems to be heading in this direction when she distinguishes between the general theory of relativity on the one hand and some more restricted theory of gravity supported by Eddington's experiments on the other.) Some have pushed this view to a point where only experimental laws are to be taken as making testable claims about the way the world is. High-level theory is seen as playing some kind of organising or heuristic role rather than making claims about the way the world is. These kinds of considerations point us in the direction of the issues discussed in the final two chapters.

Appendix: happy meetings of theory and experiment

Many agree that the merit of a theory is demonstrated by the extent to which it survives severe tests. However, there is a wide class of cases of confirmation in science that do not fit readily into this picture, unless great care is taken in characterising severity of tests. The cases I have in mind involve significant matches between theory and observation in circumstances where a lack of match would not tell against the theory. The idea is best brought out with examples.

One common kind of situation in science involves making a novel prediction from a theory in conjunction with some complicated and perhaps dubious auxiliary assumptions. When that prediction is confirmed, it is reasonable to suppose that the theory gains significant support. On the other hand, if it is not confirmed, the problem could as well lie with the auxiliary hypotheses as with the theory. Consequently, it might appear that testing the prediction did not constitute a severe test of the theory. The theory receives significant support when the prediction is confirmed nevertheless. Neil Thomason, (1994 and 1998) has developed this point in some detail. A nice example is the following. The Copernican theory predicts that Venus should exhibit phases like the moon that are correlated with its change in apparent size in a specified way, *provided it is assumed that Venus is opaque.* From a historical point of view, as Copernicus and Galileo both stated quite explicitly, the claim I have italicised was very much an open question. When, using his telescope, Galileo observed the phases of Venus, varying with the relative positions of earth, the sun and Venus and with the apparent size of Venus, in precisely the way predicted by the Copernican theory in conjunction with the assumption that Venus is opaque, this, quite reasonably, was taken as evidence offering strong support to the theory (and the auxiliary assumption). Had the phases not been observed, this could have been blamed on the auxiliary assumption as much as on the theory, so the exercise, in a sense, did not constitute a particularly severe test of the Copernican system.

A related, and quite common, situation involves exploring a theory in a messy situation where the significance of observations is far from clear. Here, a detailed matching of a theoretical prediction with observation can serve to confirm both the theory and the interpretation of the observation, whereas the failure to achieve a match simply indicates that there is more work to be done. An example involves the use of the electron microscope to observe dislocations in crystals. These dislocations, imperfections in the otherwise regular arrays of atoms in crystalline solids, had been predicted on theoretical grounds in the mid-1930s to account for the strength, ductility and plasticity of solids. If crystal structures were totally regular then the forces between crystal lattices would be much too strong to allow for the known strengths and malleability of solids. In the early 1950s electron microscopes had been developed to a stage that led some to believe that crystal lattices, and dislocations, might be observable by them, although theory of electron/specimen interaction had not been developed sufficiently to give a definite prediction one way or the other. In 1956 Jim Menter (1956) and Peter Hirsch et al. (1956) produced electron microscope images that they identified as exhibiting dislocations. Some of the ways they justified this interpretation of the complicated images are very much in line with the techniques highlighted by the new experimentalists. For example, the effects of practical intervention, such as bending the crystal, were observed to be in accordance with the assumption that the images were indeed of crystal lattices, while the effects of different physical processes, such as X-ray and electron diffraction, were shown to give mutually supportive results. More to the point that I am making here, however, is the extent to which matches between theory and observation served to confirm both. Menter, for instance, applied Abbe's theory of the microscope to the formation of electron images by crystals, and took the significant match between his predictions and the observed patterns to confirm his theory and also the interpretation of the images as images of crystal lattices. Hirsch observed his dislocations to move in just the way predicted by the prevailing theory of dislocations, and took this to confirm both that theory and the fact that his images represented dislocations. In these cases, the happy match between theory and observation constituted significant support for the theory. On the other hand, the experimental situations were sufficiently messy and ill-understood to permit plenty of explanations for failure other than implicating the theory of dislocations under test. I suggest the kind of situation I have described here can be expected to occur commonly in experimental science.

Deborah Mayo's characterisation of severity is able to accommodate these examples.[1] She will ask whether the confirmations would have been likely to occur if the theory were false. Both in the case of my Copernican example and the dislocations example, the answer is that they would be very unlikely to occur. Consequently, in each case the relevant theories received significant support from the observed coincidences between theoretical prediction and observation. Mayo's conception of severity is in line with scientific practice.

Further reading

Hacking (1983) was a pioneering work in the new experimentalism. Other works in that category are Franklin(1986) and Franklin (1990), Galison (1987) and Galison (1997) and Gooding (1990). A summary of the position is given by Ackermann (1989). The most sophisticated philosophical defence of the position is Mayo (1996).

CHAPTER 14

Why should the world obey laws?

Introduction

In the foregoing chapters we have been concerned with *epistemological* questions, that is, questions concerning how scientific knowledge is vindicated by appeal to evidence, and the nature of that evidence. In this and the next chapter we turn to *ontological* questions, questions about the kinds of things there are in the world. What kinds of entities are assumed or shown to exist in the world by modern science? Part of an answer to that question has been taken for granted in this book up until now. It has been taken for granted that there are such things as laws which govern the behaviour of the world and which it is the business of science to discover. This chapter is concerned with what kinds of entities these laws are.

The idea that the world is governed by laws that it is the business of science to discover is commonplace. However, the question of what this idea amounts to is far from being unproblematic. A fundamental problem was highlighted by Robert Boyle in the seventeenth century. The notion of a law originates in the social sphere where it makes straightforward sense. Society's laws are obeyed or not obeyed by individuals who can comprehend the laws and the consequences of violating them. But once laws are understood in this natural way, how can it be said that material systems in nature obey laws? For they can hardly be said to be in a position to comprehend the laws they are meant to obey, and, in any case, a fundamental law as it applies in science is supposed to be exceptionless, so there is no correlate to an individual's violating a social law and taking the consequences. What is it that makes matter conform to laws? This is a reasonable and straightforward question, it would appear, and yet it is not one that is easily answered. I take it that Boyle's answer, namely, that God makes matter behave in accordance with the laws He has ordained, leaves a lot to be desired from a modern point of view. Let us see if we can do better.

Laws as regularities

One common response to the question 'What makes matter behave in accordance with laws?' is to deny its legitimacy. The line of reasoning involved here was forcefully expressed by the philosopher David Hume, and has been influential ever since. From the Humean standpoint it is a mistake to assume that lawlike behaviour is caused by anything. Indeed, the whole idea of causation in nature is brought into question. The reasoning goes like this. When, for example, two billiard balls collide, we can observe their motions immediately before and immediately after collision, and we may be able to discern a regular way in which the speeds after impact are connected to the speeds before impact, but what we never see is something in addition to this, which can be identified with the causal effect of the one ball on the other. From this point of view causation is nothing other than regular connection, and laws take the form 'Events of type A are invariably accompanied or followed by events of type B'. For instance, Galileo's law of fall would take the form 'Whenever a heavy object is released near the earth's surface it falls to the ground with a uniform acceleration'. This is the so-called regularity view of laws. Nothing makes matter behave in accordance with laws because laws are nothing other than de facto regularities between events.

A standard, and telling, set of objections to the regularity view of laws involves the claim that it does not distinguish between accidental and lawlike regularities. Popper gives the example 'no moa lives beyond fifty years' as an example. It may well be the case that no moa, a species now extinct, ever lived beyond fifty years, but some might well have done so had the environmental conditions been more favourable, and for this reason we are inclined to discount the generalisation as a law of nature. But it qualifies as a law on the ground that it is an exceptionless regularity. It may well be the case that whenever the factory hooter sounds at the end of the working day in Manchester the workers down tools in London, but even if there are no exceptions to this generalisation, it hardly qualifies as a law of nature. Examples of this kind abound, and they suggest that there is something more to a law of nature than mere regularity. Another difficulty with the regularity view is that it fails to identify the direction of causal dependency. There is a regular connection between instances of smoking and lung cancer, but this is because smoking causes lung cancer, not the reverse. That is why we can hope to decrease the occurrence of cancer by eliminating smoking, but cannot hope to combat smoking by finding a cure for cancer. A regularity exhibited by events is not a sufficient condition for the regularity to constitute a law for there is more to lawlike behaviour than mere regularity.

Apart from difficulties with the idea that regularities are a sufficient condition for a law, straightforward considerations about laws as they figure in science strongly suggest that regularity is not a necessary condition either. If the view that laws describe exceptionless regular connections between events is taken seriously, then none of the claims typically taken to be scientific laws would qualify. Galileo's law of fall, mentioned above, is a case in point. Autumn leaves rarely fall to the ground with a uniform acceleration. On an unqualified regularity view this would make the law false. In a similar fashion Archimedes' principle, which claims in part that objects denser than water sink, is refuted by floating needles. If laws are taken to be exceptionless regularities, then it is very difficult to find a serious candidate for a law for want of the appropriate regularities. More to the point, most if not all of the generalities taken to be laws within science fail to qualify.

From the point of view of scientific practice, and commonsense for that matter, there is a ready response to these observations. After all, it is well understood why autumn leaves do not fall to the ground in a regular fashion. They are influenced by draughts and air-resistance which act as a disturbing influence, just as the sinking of a needle can be inhibited by surface tension. It is because physical processes are hindered by disturbing influences that physical laws characterising those processes need to be tested in contrived experimental circumstances in which the hindrances are eliminated or controlled. The regularities of relevance to science, and which are indications of lawlike behaviour, are typically the hard-won results of detailed experimentation. Think, for example, of the lengths to which Henry Cavendish had to go to get attracting spheres to exhibit the inverse square law of attraction and how J. J. Thomson eventually succeeded, where Hertz had failed, to exhibit the regular deflection of moving electrons in an electric field.

An obvious response that the defender of the regularity view of laws can give to these observations is to restate that view in a conditional form. Laws can be formulated in the form 'events of type A are regularly followed, or accompanied, by events of type B provided disturbing factors are not present'. So Galileo's law of fall becomes 'heavy objects fall to the ground with a uniform acceleration provided they do not encounter a variable resistance or are not deflected by winds or other disturbing factors'. The phrase 'other disturbing factors' is indicative of a general problem concerning how a precise statement of the conditions to be satisfied for a law to apply can be formulated. But I will leave that difficulty aside, because I suggest there is a much more fundamental one facing the regularity view here. If we accept the characterisation of laws as regularities stated in conditional form, then we must accept that laws only apply when those conditions are satisfied. Since the satisfaction of the appropriate conditions will normally only obtain in special experimental set-ups, we are forced to conclude that scientific laws generally apply only within experimental situations and not outside of them. Galileo's law of fall will be considered to apply only when heavy objects are dropped in situations where air resistance and the like have been removed. So autumn leaves are not subject to Galileo's law of fall, according to this revised version of the regularity view. Does this not clash with our intuition? Do we not wish to say that an autumn leaf is governed by the law of fall, but is also governed by the laws governing air-resistance and aerodynamics as well, so that the resulting fall is the complicated result of the various laws acting in conjunction? Because the regularity view, in its conditional form, restricts the applicability of laws to those experimental situations where the appropriate conditions are met, it is incapable of saying anything about what happens outside of those conditions. On this view, science is incapable of saying why autumn leaves usually end up on the ground!

The difficulty here echoes a problem that arises if the new experimentalism is taken as exhausting what can be said of scientific knowledge. For, as we saw in the previous chapter, although it may well be the case that the new experimentalism can capture a strong sense in which the progress of science can be understood as a steady accumulation of experimental knowledge, to leave it at that leaves us with no account of how knowledge arrived at inside experimental situations can be transported outside of those situations and used elsewhere. How are we to explain the engineer's use of physics, the use of radioactive dating in historical geology or the application of Newton's theory to the motion of comets? If scientific laws are assumed to apply outside, as well as inside, of experimental situations then laws cannot be identified with the regularities that are achievable in experimental situations. The regularity view of laws will not do.

Laws as characterisations of powers or dispositions

There is a straightforward way out of the problems with the idea of a law that we have so far discussed. It involves taking seriously what is implicit in much commonsense as well as science, namely, that the material world is active. Things happen in the world of their own accord, and they happen because entities in the world possess the capacity or power or disposition or tendency to act or behave in the way that they do. Balls bounce because they are elastic. Warnings on containers that declare the contents to be poisonous or inflammable or explosive tell us what the contents are capable of doing or how they are inclined to act. Specifying the mass and charge of an electron indicates how it will respond to electric and magnetic fields. An important element of what a thing is, is what it is capable of doing or becoming. We need to characterise things in terms of their potential as well as their actual being, as Aristotle correctly observed. Just as the ability to grow into an oak tree is an important part of what it is to be an acorn, so the capacity to attract unlike and repel like charges, and to radiate when accelerating, is an important part of what it is to be an electron. We experiment on systems to find out how they are disposed to behave.

Once we admit such things as dispositions, tendencies, powers and capacities into our characterisation of material systems, then laws of nature can be taken as characterising those dispositions, tendencies, powers or capacities. Galileo's law of fall describes the disposition heavy objects possess to fall to the ground with a uniform acceleration and Newton's law of gravitation describes the power of attraction between massive bodies. Once we interpret laws in this way, we need no longer expect laws to describe sequences of happenings in the world because those happenings will typically be the result of several dispositions, tendencies, powers or capacities acting in conjunction in complex ways. The fact that the tendency of a leaf to fall in accordance with Galileo's law is swamped by the effect of the wind is no reason in itself to doubt that that tendency continued to act on the leaf in accordance with the law. From this point of view, we can readily understand why experiment is necessary to glean information relevant for the identification of a law. The tendencies corresponding to the law under investigation need to be separated from other tendencies, and this separation requires the appropriate practical intervention to bring it about. Given the irregularities of ocean beds and the attraction of the sun and planets as well as the moon, we cannot hope to arrive at a precise account of the tides from Newton's theory plus initial conditions. Nevertheless, gravity is the major cause of the tides and there are appropriate experiments for identifying the law of gravity.

From the point of view I am advocating, causes and laws are intimately linked. Events are caused through the action of particulars that possess the power to act as causes. The gravitational attraction of the moon is the main cause of the tides, charged particles cause the ionisation responsible for the tracks in a cloud chamber and oscillating charges cause the radio waves emitted from a transmitter. Descriptions of the mode of acting of the active powers involved in such cases constitute the laws of nature. The inverse square law of gravitation describes quantitatively the power to attract possessed by massive bodies, and the laws of classical electromagnetic theory describe, among other things, the capacity of charged bodies to attract and radiate. It is the active powers at work in nature that makes laws true when they are true. We thus have a ready answer to Boyle's question. It is the powers and capacities possessed by particulars and operative when particulars interact that compel those particulars to behave in accordance with laws. Lawlike behaviour is brought about by efficient causation. Boyle faced the problem he did with laws, and needed to invoke God, just because he declined to ascribe dispositional properties to matter.

The majority of philosophers seem reluctant to accept an ontology which includes dispositions or powers as primitive. I do not understand their reluctance. Perhaps the reasons are in part historical. Powers were given a bad name by the mystical and obscure way they were employed in the magical tradition in the Renaissance, and they are alleged to have been exploited by the Aristotelians in a cavalier way under the guise of forms. Boyle's rejection of active properties in his mechanical philosophy can be seen as a reaction, and perhaps an overreaction, to the excesses of those traditions, as well as being motivated by theological concerns. However, there need be nothing mysterious or epistemologically suspect about invoking powers, tendencies and the like. Claims concerning them can be subject to stringent empirical tests to as great an extent as any other kind of claim. What is more, however much philosophers may be averse to dispositional properties, scientists systematically invoke them and their work would be incapacitated without them. It is significant to note in this respect that Boyle, in his experimental science as opposed to his mechanical philosophy, freely employed dispositional properties such as acidity and the spring of the air. Elasticity in various forms was an embarrassment to the seventeenth-century mechanical philosophers. Hobbes complained that Boyle's attribution of elasticity to air was equivalent to the admission that air could move itself. Boyle and other seventeenth-century scientists continued to employ the concept of elasticity, and never succeeded in explaining it away by reference to non-dispositional properties. Nor has anyone succeeded since. I do not understand what grounds philosophers have for questioning, or feeling the need to explain away, this common, indeed ubiquitous, usage by scientists of dispositional properties.

The view that laws characterise the dispositions, powers, capacities or tendencies of things has the merit that it acknowledges at the outset what is implicit in all scientific practice, namely, that nature is active. It makes clear what makes systems behave in accordance with laws, and it links laws with causation in a natural way. It also offers a ready solution to the problem, encountered in the previous chapter, concerning the transportability of knowledge acquired in experimental situations beyond those situations. Once the assumption is made that entities in the world are what they are by virtue of the powers and capacities that they possess, and I claim that that assumption is implicit in scientific practice as well as everyday life, then the laws describing those powers and capacities, identified in experimental situations, can be presumed to apply outside of those situations too. Nevertheless, I cannot leave things here with a good conscience, because there are important laws of science that are difficult to fit into this scheme.

Thermodynamic and conservation laws

Let us refer to the view I have outlined and defended in the previous paragraph, which understands laws as characterising causal powers, as the causal view of laws. There are important laws in physics that do not fit well into this scheme. The first and second laws of thermodynamics do not and nor do a range of conservation laws in fundamental particle physics. The first law of thermodynamics asserts that the energy of an isolated system is constant. The second law, which asserts that the entropy of an isolated system cannot decrease, has consequences such as ensuring that heat flows from hot to cold bodies and not the other way round and ruling out the possibility of extracting heat energy from the sea and putting it to useful work, where the only price paid for the work is a decrease in temperature of the sea. A machine that succeeded in doing this would be a perpetual motion machine of the second kind, distinct from a machine that results in a net increase in energy, which is a perpetual motion machine of the first kind. The first law of thermodynamics rules out perpetual motion machines of the first kind and the second law rules out perpetual motion machines of the second kind. These quite general laws have consequences for the behaviour of physical systems, and can be used to predict their behaviour, quite independently of the details of the causal processes at work. That is why it is not possible to construe these laws as causal laws.

Let me give an example that illustrates my point. If ice is subjected to pressures higher than normal atmospheric pressure its melting point is lowered. This is why a wire from which weights are suspended will cut its way through a block of ice. The explanation of this at the molecular level is far from straightforward and a precise, detailed account is probably not available. Since pressure tends to push molecules closer together, one might expect the forces of attraction between them to increase under such circumstances, leading to an increase in the thermal energy necessary to drag them apart and thus to an elevation in melting point. This is precisely what happens in a typical solid near melting point. But ice is not a typical solid. The water molecules in ice are rather loosely packed, more so than they are in the liquid state, which is why ice is less dense than water. (This is just as well, otherwise lakes and rivers would freeze from the bottom up, and would freeze in their entirety in periods of prolonged cold, thus eliminating fish and anything evolved from fish as a viable life form.) If the molecules in ice are forced closer together than normal, the force between them decreases, so less thermal energy is needed to separate them, and the melting point falls. The precise way in which the forces depend on molecular positions is complicated, depending on fine quantum mechanical detail involving exchange as well as Coulomb forces, and is not known with precision.

Given the above complications, it may come as a surprise that James Thomson was able to predict the depression of the freezing point of water with pressure in 1849 thereby anticipating the empirical discovery of the phenomenon. All he needed for his derivation were the laws of thermodynamics plus the empirically known fact that water is denser than ice. Thomson devised, in thought, a cyclic process that involved extracting heat from water at 0°C and converting it into ice at 0°C. It seemed as if this engine provided a means of extracting heat from water and converting all of it into the work done by the expansion involved, thus comprising a perpetual motion machine of the second kind, ruled out by the second law of thermodynamics. Thomson realised that this unacceptable conclusion could be blocked by assuming the freezing point to be lowered by an increase in pressure.

The feature of this case that I wish to highlight is that Thomson's prediction was made in ignorance of the details of the causal process at the molecular level. A characteristic feature, and a major strength, of thermodynamics is that it applies at the macroscopic level whatever the details of the underlying causal process. It is precisely this feature of the laws of thermodynamics that prevents them being construed as causal laws.

The difficulties for the causal view do not stop here. The behaviour of a mechanical system can be understood and predicted by specifying the forces on each component of the system and using Newton's laws to trace the development of the system. Within this approach Newton's laws can readily be interpreted as causal laws describing the disposition of objects to exert and respond to specified forces. However, this is not the only way of dealing with mechanical systems. The laws of mechanics can also be written in a form that takes energy, rather than forces, as the starting point. In the Hamiltonian and Lagrangian formulations of mechanics, where this approach is adopted, what is required is expressions for the potential and kinetic energy of a system as a function of whatever coordinates are necessary to fix them. The evolution of a system can then be completely specified by feeding these expressions into the Hamiltonian or Lagrangian equations of motion. This can be done without a detailed knowledge of the causal processes at work.

James Clerk Maxwell (1965, vol. 2, pp. 783–4), who attempted to cast his electromagnetic theory in Lagrangian form, illustrated this point in a characteristically vivid way. We imagine a belfry in which a complicated piece of machinery is driven by bell ropes that drop to the bell ringer's room below. We assume the number of ropes to be equal to the number of degrees of freedom of the system. The potential and kinetic energy of the system as a function of the position and velocity of the ropes can be determined by experiments done with the ropes. Once we have these functions we can write down Lagrange's equations for the system. It is then possible, given the positions and velocities of the ropes at any one instant, to derive their positions and velocities at any other instant. We can do this without needing to know the details of the causal story of what is happening in the belfry. Lagrange's equations do not state causal laws.

It might be objected that these observations about the Lagrangian formulation of mechanics do not constitute a serious counter example to the causal view of laws. It might be pointed out, for example, that, although a Lagrangian treatment of the mechanism in the belfry can work as well as it does by ignoring the detailed causal story of the mechanism in the belfry, there is such a story to be had that can be formulated in Newtonian, and hence causal, terms once appropriate empirical access to the belfry is gained. After all, it might be observed, Lagrange's equations can be derived from Newton's.

This last claim is no longer true (if it ever was). In modern physics Lagrange's equations are interpreted in a more general way than the version of those equations that can be derived from Newton's laws. The energies involved are interpreted in a general way that includes all kinds of energy, not just energy arising from the motion of massive bodies under the influence of forces. For instance, the Lagrangian formulation can accommodate electromagnetic energy, which includes velocity-dependent potential energies and necessitates such things as the electromagnetic momentum of a field, which is a momentum different from that corresponding to a mass times velocity. When pushed to the limit in modern physics, these Lagrangian (or related Hamiltonian) formulations are not such that they can be replaced by the causal accounts that underlie them. For instance, the various conservation principles, such as conservation of charge and parity, intimately connected with symmetries in the Lagrangian function of the energies, are not explicable by reference to some underlying process.

The outcome of all this can be summarised as follows. A wide range of laws within physics can be understood as causal laws. When this is possible, there is a ready answer to Boyle's question concerning what it is that compels physical systems to behave in accordance with laws. It is the operation of the causal powers and capacities characterised by laws that make systems obey them.

However, we have seen that there are fundamental laws in physics that cannot be construed as causal laws. In these cases there is no ready answer to Boyle's question. What makes systems behave in accordance with the law of conservation of energy? I don't know. They just do. I am not entirely comfortable with this situation, but I don't see how it can be avoided.

Further reading

For a different view of laws than the one characterised here, and for a detailed critique of the regularity view, see Armstrong (1983). The way in which experiment points towards the causal view of laws is shown in Bhaskar (1978). Cartwright (1983) casts doubt on the idea that there can be fundamental laws that are true of the world, but modifies her views to defend something more like the causal view in her 1989 text. The clash between how many philosophers characterise laws and the notion of laws employed by scientists is described with interesting examples in Christie (1994). The material of this chapter is largely derived from, and is dealt with in a little more detail in, Chalmers (1999). Another recent discussion of the nature of laws is van Fraassen (1989). Psillos (2002, pp. 122–68) is a good overall summary of the issue.

CHAPTER 15

Realism and anti-realism

Introduction

A natural assumption to make about scientific knowledge is that it tells us much about the nature of the world that goes well beyond what it appears to be like on the surface. It tells us about electrons and DNA molecules, the bending of light in gravitational fields, and even about the conditions that prevailed in the world long before there were humans to observe it. Not only does science aim to give us knowledge of such things, but it has, in the main, succeeded in doing so. Science describes not just the observable world but also the world that lies behind the appearances. This is a rough statement of *realism* with respect to science.

Why would anyone wish to deny realism? There are certainly many contemporary philosophers of science that do. One source of doubts about realism is the extent to which claims about the unobservable world must be hypothetical to the extent that they do transcend what can be firmly established on the basis of observation. Realism with respect to science is too rash, it would seem, insofar as it claims more than can reasonably be defended. These doubts can be reinforced by a historical reflection. Many theories of the past which made claims about unobservable entities did indeed turn out to be rash in this respect because they have been rejected. Newton's particle theory of light, the caloric theory of heat, and also Maxwell's electromagnetic theory, insofar as it assumed electric and magnetic fields to be states of a material ether, provide examples. Although the theoretical parts of those theories have been rejected, the anti-realist can note, those parts of them that were based on observation have been retained. Newton's observations concerning chromatic aberration and interference, calorimetry, Coulomb's law of attraction and repulsion of charged bodies and Faraday's laws of electromagnetic induction have been incorporated into modern science. The enduring part of science is that part which is based on observation and experiment. The theories are mere scaffolding which can be dispensed with once they have outlived their usefulness. This is the typical anti-realist position.

So the realist position reflects the unthinking attitude of most scientists and non-scientists, and realists will ask 'how could scientific theories involving unobservable entities such as electrons and gravitational fields be as successful as they are if they did not correctly describe the unobservable realm, at least approximately?' The anti-realist, in response, stresses the inconclusiveness of the evidence for the theoretical part of science and points out that, just as theories in the past proved successful in spite of the fact that they were not correct descriptions of reality, so it is reasonable to assume the same about contemporary ones. This is the debate that we explore in this chapter.

Global anti-realism: language, truth and reality

There is a form that the realism/anti-realism debate frequently takes in contemporary literature that I do not think is helpful, and which, in any case, is a different debate from the one I, and many others, wish to address. Readers who are unimpressed by the general and abstract terms of the discussion can safely skip this section. Global anti-realism, as I will call it, raises the question of how language of any kind, including scientific language, can engage with, or hook onto, the world. Its defenders observe that we have no way of coming face to face with reality to read off facts about it, by way of perception or in any other way. We can view the world only from our humanly generated perspectives and describe it in the language of our theories. We are forever trapped within language and cannot break out of it to describe reality 'directly' in a way that is independent of our theories. Global anti-realism denies we have access to reality generally, and not just within science.

I doubt if any serious contemporary philosopher holds that we can come face to face with reality and directly read off facts about it. I remind the reader that in this book we left any such idea behind round about chapter 2. So in that sense we are all global anti-realists, but that is not saying much because it is such a weak thesis. It becomes a stronger thesis when this lack of direct access to reality is taken to have consequences justifying a skeptical attitude towards science and to knowledge generally. The idea seems to be that no knowledge can have any kind of privileged position as a characterisation of the world because we lack the kind of access to the world that would serve to justify this. This move is unwarranted. Although it is true that we cannot describe the world without using some conceptual framework, we can nevertheless test the adequacy of those descriptions by interacting with the world. We find out about the world not just by observing and describing it but by interacting with it. As discussed in chapter 1, the construction of, necessarily linguistically formulated, claims about the world is one thing. Their truth or falsity is another. The notion of truth is often seen as having an important bearing on the debates about realism, so a discussion of the notion is called for.

The theory of truth most conducive to the needs of a realist is the so-called correspondence theory of truth. The general idea is straightforward enough and can be illustrated in commonsense terms in a way that makes it appear almost trivial. According to the correspondence theory, a sentence is true if and only if it corresponds to the facts. The sentence 'the cat is on the mat' is true if the cat is on the mat and is false if it isn't. A sentence is true if things are as the sentence says they are and false otherwise.

One difficulty with the notion of truth is the ease with which it can lead to paradoxes. The so-called liar paradox provides an example. If I say 'I never tell the truth' then if what I have said is true then what I have said is false! Another example goes as follows. We imagine a card, on one side of which is written 'the sentence on the other side of this card is true', and on the other side is written 'the sentence written on the other side of this card is false'. A little thought will reveal the paradoxical conclusion that either of the sentences are both true and false.

The logician Alfred Tarski demonstrated how, for a reasonably simple language system, paradoxes can be avoided. The crucial step was his insistence that, when one is talking of the truth or falsity of the sentences in some language, one must carefully distinguish sentences in the language system that is being talked about, the 'object language', from sentences in the language system in which talk about the object language is carried out, the 'metalanguage'. Referring to the paradox involving the card, if we adopt Tarski's recommendation then we must decide whether each sentence on the card is in the language being talked about or in the language in which the talking is being done. If one follows the rule that each of the sentences must be in either the object or the metalanguage but not in both, then neither sentence can both refer to the other and be referred to by the other, and no paradoxes arise.

A key idea of Tarski's correspondence theory, then, is that if we are to talk about truth for the sentences of a particular language, then we need a more general language, the metalanguage, in which we can refer both to the sentences of the object language and to the facts to which those object language sentences are intended to correspond. Tarski needed to be able to show how the correspondence notion of truth can be systematically developed for all sentences within the object language in a way that avoids paradoxes. The reason that this was a technically difficult task is that for any interesting language there is an infinite number of sentences. Tarski achieved his task for languages involving a finite number of single placed predicates, that is, predicates such as 'is white' or 'is a table'. His technique involved taking as given what it means for a predicate to be satisfied by an object. Examples from everyday language sound trivial. For instance, the predicate 'is white' is satisfied by $\underline{x}$ if and only if $\underline{x}$ is white. Given this notion of satisfaction for all the predicates of a language, Tarski showed how the notion of truth can be built up from this starting point for all the sentences of the language. (To use technical terminology, taking the notion of primitive satisfaction as given, Tarski defined truth recursively.)

Tarski's result was certainly of major technical importance for mathematical logic. It had a fundamental bearing on model theory and also had ramifications for proof theory. But these are matters far beyond the scope of this book. Tarski also showed how it is that contradictions can arise when truth is discussed in natural languages, and showed how those contradictions can be avoided.

But I do not think he did more than that, and Tarski himself seemed not to have thought so either. For our purposes I suggest there is nothing more to Tarski's correspondence theory than is encapsulated in the trivial sounding prescription 'snow is white' is true if and only if snow is white. That is, Tarski has shown that a commonsense idea of truth can be utilised in a way that is free from the paradoxes that were thought to threaten it. From this point of view, a scientific theory is true of the world if the world is the way the theory says it is, and false otherwise. Insofar as our discussion of realism involves a notion of truth, this is the notion of truth that I will employ.

Those keen on defending global anti-realism maintain that the correspondence theory of truth does not escape from language to describe a relationship between sentences and the world in the way it is claimed to. If I am asked what a statement such as 'the cat is on the mat' corresponds to, then unless I refuse to answer I must offer a statement in reply. I will reply 'the cat is on the mat' corresponds to the cat's being on the mat. Those who support the objection I have in mind would respond to this by saying that in giving my reply I have not characterised a relationship between a statement and the world but between a statement and another statement. That this is a misguided objection can be brought out with an analogy. If I have a map of Australia and I am asked to what the map refers, then the answer is 'Australia'. In giving this answer I am not saying that the map refers to the word 'Australia'. If I am asked what the map refers to, I have no alternative but to give a verbal reply. The map is a map of a large land mass that is *named* Australia. Neither in the case of the cat nor the map can it be sensibly said that the verbal reply involves me in the claim that, in the first case, the sentence 'the cat is on the mat' and, in the second case, the map refers to something verbal. (It seems to me that, for example, Steve Woolgar's (1988) global anti-realism with respect to science involves the confusion I have tried to unravel here.) To me at least, the claim that 'the cat is on the mat' refers to a state of affairs in the world and is true if the cat is on the mat and false if it isn't is perfectly intelligible and trivially correct.

A realist will typically claim that science aims at theories that are true of the world, both observable and unobservable, where truth is interpreted as the commonsense notion of correspondence to the facts. A theory is true if the world is as the theory says it is and false otherwise. In the case of cats on mats, the truth of statements can be fairly straightforwardly established. In the case of scientific theories this is far from being the case. I repeat, the brand of realism I wish to explore does not involve the claim that one can come face to face with reality and read off which facts are true and which are false.

The traditional debate between realists and anti-realists with respect to science concerns the issue of whether scientific theories should be taken as candidates for the truth in an unrestricted sense, or whether they should be taken as making claims about the observable world only. So both sides see science aiming at truth in some sense (a sense which I will interpret as correspondence of the kind discussed above). So neither side of the debate supports global anti-realism. So let us leave global anti-realism behind and get down to the serious business.

Anti-realism

The anti-realist maintains that the content of a scientific theory involves nothing more than the set of claims that can be substantiated by observation and experiment. Many anti-realists can usefully be called, and often are called, *instrumentalists*. For them theories are nothing more than useful instruments for helping us to correlate and predict the results of observation and experiment. Theories are not appropriately interpreted as being true or false. Henri Poincaré (1952, p. 211) exemplified this position when he compared theories to a library catalogue. Catalogues can be appraised for their usefulness, but it would be wrong-headed to think of them as true or false. So it is with theories for the instrumentalist. The latter will demand of theories that they be general (bringing under their umbrella a wide range of kinds of observation) and simple, as well as the main requirement, that they be compatible with observation and experiment. Bas van Fraassen (1980) is a contemporary anti-realist who is not an instrumentalist insofar as he thinks that theories are indeed true or false.

However, he regards their truth or falsity as beside the point as far as science is concerned. For him the merit of a theory is to be judged in terms of its generality and simplicity and the extent to which it is borne out by observation and leads to new kinds of observation. Van Fraassen calls his position 'constructive empiricism'. An advocate of the new experimentalism who sees the growth of science in terms of the growth of controllable scientific effects and nothing more would qualify as an anti-realist in the sense I am using that term.

A motivation underlying anti-realism seems to be the desire to restrict science to those claims that can be justified by scientific means, and so avoid unjustifiable speculation. Anti-realists can point to the history of science to substantiate their claim that the theoretical part of science does not qualify as securely established. Not only have theories of the past been rejected as false, but many of the entities postulated by them are no longer believed to exist. Newton's corpuscular theory of light was an accepted part of physics for over a hundred years. Not only is it now regarded as false, but there are no such things as the corpuscles that Newton's optics implied. The ether that was centrally involved in nineteenth-century wave optics and electromagnetic theory has been similarly discarded, and a key idea in Maxwell's theory, that electric charge is nothing other than a discontinuity in a strain in the ether, is now regarded as plain wrong. However, the anti-realist will insist that, although these theories proved to be untrue, there is no denying the positive role they played in helping to order, and indeed to discover, observable phenomena. After all, it was Maxwell's speculations about electromagnetism as representing states of an ether that led him to an electromagnetic theory of light and was eventually to lead to the discovery of radio waves. In the light of this, it seems plausible to evaluate theories solely in terms of their ability to order and predict observable phenomena. As such, the theories themselves can be discarded when they have outlived their usefulness, and the observational and experimental discoveries to which they have led can be retained. Just as past theories and the unobservable entities employed by them have been discarded, so we can expect our present ones to be. They are simply scaffolding to help erect the structure of observational and experimental knowledge, and they can be rejected once they have done their job.

Some standard objections and the anti-realist response

The anti-realist presupposes a distinction between knowledge at the observational level, which is regarded as securely established, and theoretical knowledge, which cannot be securely established and is best seen as a heuristic aid. The discussion of the theory-dependence and fallibility of observation and experiment in the early chapters of the book poses problems for this view, at least on the surface. If observation statements and experimental results are regarded as acceptable to the extent that they can survive tests, but are liable to be replaced in the future in the light of new, more discerning tests, then this opens the way for the realist to treat theories in exactly the same way, and to deny that there is a fundamental or sharp distinction between observational and theoretical knowledge of the kind that the anti-realist bases his or her position on.

Let us engage with this issue at the level of experiment rather than mere observation. Here the anti-realist need not deny that theory plays a role in the discovery of new experimental effects. He or she can stress, however, as I did in the chapter on the new experimentalism, that new experimental effects can be appreciated and manipulated in a way that is independent of theory, and this experimental knowledge does not get lost when there is a radical theory change. I gave Faraday's discovery of the electric motor and Hertz's production of radio waves as examples. Cases such as these can be deployed in a way that gives credence to the anti-realist's position. Whether all experimental results as they figure in science can be construed as theory-independent in this kind of way is disputable, however. Let me crystallise the problem by invoking again my story about the use of the electron microscope to investigate dislocations in crystals. Some aspects of the early work can aid the anti-realist. The validity of the observations of dislocations was established by various manipulations and cross-checks that did not rely on an appeal to a detailed theory of the electron microscope and the interaction of electron beams with crystals. However, as the work got more sophisticated, interpretations of the observable images could only be achieved and supported by the agreement between fine detail and the predictions of theory. There is no denying that knowledge of dislocations has been of immense practical importance for understanding the strengths of materials and many other properties of solids. What an anti-realist needs to be able to do is show how the experimentally useful part of that knowledge can be formulated and vindicated in a way that is independent of theory. The issue is discussed in more detail in Chalmers (2003).

Another standard objection to anti-realism concerns the predictive success of theories. How can it be, so the objection goes, that theories are so predictively successful if they are not at least approximately true? The argument seems to have particular force in those cases where a theory leads to the discovery of a new kind of phenomenon. How can Einstein's theory of general relativity be considered as a mere calculating device given that it successfully predicted the bending of light rays by the sun? How can it be seriously maintained that the structures attributed to organic molecules were mere instruments when those structures can now be witnessed 'directly' with electron microscopes?

The anti-realist can respond as follows. They can certainly agree that theories can lead to the discovery of new phenomena. Indeed, this is one of the desiderata they themselves place on a good theory. (Remember, it is not part of the anti-realist's position that there is no place for theory in science. It is the status of theory that is in question.) However, the fact that a theory is productive in this respect need be no indication that it is true. This is evident from the fact that theories of the past have proved successful in this respect even though, from a modern point of view, they cannot be regarded as true. Fresnel's theory of light as waves in an elastic ether successfully predicted the bright spot discovered by Arago and Maxwell's speculations about the displacement of the ether led to the prediction of radio waves. The realist regards Newton's theory as false in the light of Einstein's theory and quantum mechanics. And yet Newton's theory had over two centuries of predictive success to its credit before it was eventually refuted. So doesn't history force the realist to admit that predictive success is not a necessary indication of truth?

There is an important historical episode in the history of science that has been used in attempts to discredit anti-realism. It involves the Copernican revolution. As we have seen, Copernicus and his followers faced problems defending their claim that the earth moves. One response to those problems was to take an anti-realist stance with respect to that theory, deny that it be taken literally as describing true motions, and demand merely that it be compatible with astronomical observations. A clear expression of this view was formulated by Osiander in the Preface that he wrote for Copernicus's main work, *The Revolutions of the Heavenly Spheres*. He wrote,

> . . . it is the duty of an astronomer to compose the history of the celestial motions through careful and skillful observation. Then turning to the causes of these motions or hypotheses about them, he must conceive and devise, since he cannot in any way attain to the true causes, such hypotheses as, being assumed, enable the motions to be calculated correctly from the principles of geometry, for the future as well as the past. The present author [Copernicus] has performed both these duties excellently. For the hypotheses need not be true nor even probable; if they provide a calculus consistent with observation that alone is sufficient. (Rosen, 1962, p. 125)

By taking this stance, Osiander and like-minded astronomers were removed from the need to face up to the difficulties posed by the Copernican theory, especially those stemming from the claim that the earth moves. Realists such as Copernicus and Galileo, however, were forced to try to face up to those difficulties and attempt to remove them. In Galileo's case this led to major advances in mechanics. The moral that the realist wishes to draw from this is that anti-realism is unproductive because difficult questions, which demand a solution from a realist perspective, are swept under the carpet by anti-realists.

The anti-realist can respond that this example is a caricature of the anti-realist position. Among the demands that an anti-realist makes of theories is the insistence that theories be general and unified – that they embrace a wide range of phenomena. From this perspective, the anti-realist must seek to embrace astronomy and mechanics under one theoretical framework, and so would be just as motivated to tackle the mechanical problems associated with the Copernican theory as the realist. It is ironic in this connection that a prominent anti-realist, Pierre Duhem (1969), in his book *To Save the Phenomena*, chose the example of the Copernican revolution to support his case!

Scientific realism and conjectural realism

I begin by stating realism in a very strong form, to which some have given the name 'scientific realism'. According to scientific realism, science aims at true statements about what there is in the world and how it behaves, at all levels, not just at the level of observation. What is more, it is claimed that science has made progress towards this aim, insofar as it has arrived at theories that are at least approximately true and discovered at least some of what there is. So, for example, science has discovered that there are such things as electrons and black holes, and, although earlier theories about such entities have been improved upon, those earlier theories were approximately true, as can be shown by deriving them as approximations to current theory. We cannot know that our current theories are true, but they are truer than earlier theories, and will retain at least approximate truth when they are replaced by something more accurate in the future. These claims are regarded by the scientific realist as on a par with scientific claims themselves. It is claimed that scientific realism is the best explanation of the success of science and can be tested against the history of science and contemporary science in much the same way as scientific theories are tested against the world. It is the claim about the testability of realism against the history of science that is seen as warranting the naming of this brand of realism 'scientific'. Richard Boyd (1984) has given a clear statement of scientific realism of the kind I have summarised here.

A key problem for this strong version of realism stems from the history of science and the extent to which that history reveals science to be fallible and revisable. The history of optics provides the strongest example. Optics has undergone fundamental changes in its progress from Newton's corpuscular theory through to modern times. According to Newton, light consists of beams of material corpuscles. Fresnel's theory, which replaced it, construed light as a transverse wave in an all-pervasive elastic ether. Maxwell's electromagnetic theory of light reinterpreted these waves as involving fluctuating electric and magnetic fields, although the idea that those fields were states of an ether was retained. By early in the twentieth century the ether had been eliminated leaving the fields as entities in their own right. It soon became necessary to supplement the wave character of light with a particle aspect by introducing photons. I take it that realists and anti-realists alike consider this series of theories to have been progressive from beginning to end. But how can this progress be reconciled with the scientific realist's strictures? How can this series of theories be construed as moving towards better and better approximations to a characterisation of what there is in the world, when what is in evidence is a drastic fluctuation? First light is characterised in terms of particles, then waves in an elastic medium, then as fluctuating fields-in-themselves and then as photons.

Admittedly, there are other examples that seem to fit the realist picture better. The history of the electron is a case in point. When it was first discovered in the form of cathode rays towards the end of the nineteenth century, it was construed as simply a tiny particle with a small mass and an electric charge. Bohr needed to qualify this picture in his early version of a quantum theory of the atom, in which electrons orbited a central positive nucleus but without radiating, as circling charged particles would be expected to do. They are now regarded as quantum mechanical entities that have a half integral spin, can behave like waves in appropriate circumstances and obey Fermi-Dirac rather than classical statistics. It is reasonable to suppose that throughout this history it is the same electrons that are being referred to and experimented on, but that we have steadily improved and corrected our knowledge of them, so that it is reasonable to see the sequence of theories about electrons as approaching truth. Ian Hacking (1983) has indicated a way in which the realist position can be strengthened from this kind of perspective. He argues that the anti-realists place an inappropriately strong emphasis on what can and what cannot be observed and pay insufficient attention to what can be practically manipulated in science. He argues that entities in science can be shown to be real once they can be practically manipulated in a controlled way and used to bring about effects in something else. Beams of positrons can be produced and trained on targets to bring about effects in a controlled way, so how can they not be real, in spite of the fact that they cannot be directly observed? If you can spray them, says Hacking, then they are real (p. 23). If this criterion for judging what is real is adopted, then perhaps my example concerning particles of light and the ether need not tell against realism, because those entities were never established as real by practically manipulating them.

There are realists who regard scientific realism as too strong and attempt to weaken it in various ways. The brand of realism advocated by Popper and his followers is of that kind and can be referred to as conjectural realism. The conjectural realist stresses the fallibility of our knowledge, and is well aware that theories of the past, together with their claims about the kinds of entities there are in the world, have been falsified and replaced by superior theories that construe the world quite differently. There is no knowing which of our current theories might suffer a similar fate. So the conjectural realist will not claim that our current theories have been shown to be approximately true, nor that they have conclusively identified some of the kinds of things there are in the world. The conjectural realist will not rule out the possibility that the electron might suffer the same fate as the ether. Nevertheless, it is still maintained that it is the *aim* of science to discover the truth about what really exists and theories are to be appraised for the extent to which they can be said to fulfil that aim. The conjectural realist will say that the very fact that we can declare past theories to be false indicates that we have a clear idea of the ideal that those past theories have fallen short of.

Although conjectural realists will insist that their position is the most fruitful one to adopt in science, they will stop short of describing their position as scientific. Scientific realists claim that their position can be tested against the history of science and can explain the success of science. The conjectural realist regards this as too ambitious. Before a theory in science can be accepted as an explanation for a range of phenomena it can reasonably be demanded that there be some independent evidence for the theory, independent, that is, of the phenomena to be explained. As John Worrall (1989b, p. 102) has pointed out, there is no question of scientific realism living up to this demand since there is no question of there being evidence independent of the history of science which scientific realism is meant to explain. The general point is that it is difficult to see how scientific realism can be confirmed by the historical evidence once one takes seriously the kind of stringent demands made within science itself concerning what counts as a significant confirmation. Conjectural realism is seen as a philosophical, rather than scientific, position by the conjectural realist, to be defended in terms of the philosophical problems it can solve.

A major problem with conjectural realism is the weakness of its claims. It does not claim that current theories can be known to be true or approximately true nor does it claim that science has conclusively discovered some of the things that there are in the world. It simply claims that science aims to achieve such things, and that there are ways of recognising when science falls short of this aim. The conjectural realist has to admit that even if true theories and true characterisations of what there is were arrived at in science there would be no way of knowing it. It might well be asked what differences there are between this view and that of the most sophisticated anti-realist when it comes to an understanding and appraisal of current or past science.

Idealisation

A standard objection to realism, raised by Duhem (1962, p. 175), for example, is that theories cannot be taken as literal descriptions of reality because theoretical descriptions are idealised in a way that the world is not. We will all recall that the science we learnt at school involved such things as frictionless planes, point masses and inextensible strings and we all know that there are no items in the world that match these descriptions. Nor should it be thought that these are simplifications introduced only in elementary texts, with more complicated descriptions characterising the real state of affairs introduced later in more advanced science. Newtonian science inevitably makes approximations in astronomy, for example, treating the planets as point masses or homogeneous spheres and the like. When quantum mechanics is used to derive the properties of the hydrogen atom, such as its characteristic spectra, it is treated as a negatively charged electron moving in the vicinity of a positively charged proton, isolated from its surroundings. No real hydrogen atom is ever isolated from its surroundings. Carnot cycles and ideal gases are other idealisations that play a crucial role in science without there being counterparts to them in the real world. Finally, we note that from a realist perspective, the parameters that are taken to characterise systems in the world, such as the position and velocity of a planet or the charge on the electron, are treated as indefinitely precise when manipulated by exact mathematical equations, whereas experimental measurements are always accompanied by some margin of error, so that a measured quantity will be denoted as $x \pm dx$, where dx represents the margin of error. The general idea, then, is that in various ways, theoretical descriptions are idealisations that cannot correspond to real-world situations.

My own view is that idealisations in science do not pose the difficulties for realism that they are often thought to do. As far as the undoubted inaccuracy of all experimental measurements is concerned, it does not follow from this that the quantities measured do not possess precise values. I would argue, for example, that in physics we have strong evidence for the claim that the charge on every electron is absolutely identical, in spite of the inaccuracy of measurements of that charge. Many macroscopic properties, such as the conductivity of metals and the spectra of gases, depend on the way electrons, because of the strong sense in which they are identical, obey Fermi-Dirac rather than classical, Boltzmann statistics. This example is not likely to impress the anti-realist who regards the electron as a theoretical fiction, but, like Hacking, it seems to me that the experimental manipulation of electrons that is now commonplace makes an anti-realist attitude with respect to them extremely implausible.

Idealisation can be viewed in an instructive way in the light of the discussion of the nature of laws in the previous chapter. There it was suggested that a common class of laws describes the powers and tendencies of particulars to behave or act in certain ways. It was stressed that observable sequences of events should not be expected to reflect the orderly action of these powers and tendencies because the systems in which they operate will typically be complex and involve the simultaneous operation of other powers and tendencies. So, for instance, however accurate we attempt to make an experiment designed to measure the deflection of cathode rays in a discharge tube, we will never be able to completely eliminate the effect of the gravitational attraction on the electrons due to nearby masses, the effect of the earth's magnetic field, and so on. To the extent that it is accepted that the causal account of laws is able to make sense of laws in science where the regularity view fails, then this requires us to view laws as describing causal powers that act behind the appearances, combining with other powers to yield resulting events or sequences of events that may be observable. That is, the causal account of laws is a realist account. The anti-realist seems obliged to capture the functioning of laws in science with some version of the regularity view. We discussed the difficulties they face in the previous chapter.

Unrepresentative realism or structural realism

If we take the most sophisticated versions of realism and anti-realism, then each seems to have a major point in its favour. The realist can point to the predictive success of scientific theories, and can ask, how can this success be explained if theories are mere calculating devices? The anti-realist can counter by pointing out that past scientific theories were predictively successful even though the realist is forced to characterise them as false. This dramatic turnover in theories is the key point in favour of the anti-realist. Is there a position that manages to capture the best of both worlds? In the past I have attempted to do so with a position I called unrepresentative realism. That view has similarities with a position developed by John Worrall (1989b) which he calls structural realism. My phrase has not caught on. Worrall seems to be having more luck.

The history of optics provided us with the most problematic

example from the realist point of view, because there we see undoubtedly successful theories being overthrown with an accompanying change in the understanding of the kind of thing that light is. So let us concentrate on this problem case and see to what extent a realist view can be salvaged. Popperian realists, in their zeal to combat positivist or inductivist understandings of science, point to the falsification of previously well-confirmed theories to bolster their case that scientific knowledge remains fallible however much positive evidence there is in its favour. In this spirit they will insist, for example, that Fresnel's wave theory of light has been shown to be false. (There is no elastic ether and the wave theory cannot handle phenomena such as the photoelectric effect, where light exhibits its particle-like nature.) But is it helpful or accurate to dismiss Fresnel's theory as simply false? After all, there is a wide range of circumstances in which light does behave like a wave. There was more to Fresnel's theory than mere predictive success. It successfully captured something right about light in a wide range of circumstances, its wave-like structure as exhibited in those circumstances. It was because Fresnel's theory successfully captured that structure that it was predictively successful, leading to dramatically successful predictions such as the famous white spot. Worrall stresses this point by focusing on the *mathematical* structure of Fresnel's theory and points out that many of the equations that figure in Fresnel's treatment of light, such as the equations giving the detail of reflection and refraction at transparent surfaces, are retained in current theory. That is, from the point of view of the contemporary understanding of the matter, Fresnel's equations provide true, not false, descriptions of a wide range of optical phenomena, notwithstanding the fact that some of Fresnel's interpretations of the reality underlying his equations have been discarded.

So science is realist in the sense that it attempts to characterise the structure of reality, and has made steady progress insofar as it has succeeded in doing so to an increasingly accurate degree. Past scientific theories were predictively successful to the extent that they did at least approximately capture the structure of reality (so their predictive success is not an unexplained miracle), so a major problem with anti-realism is avoided. On the other hand, while science steadily progresses insofar as the structures attributed to reality are constantly refined, the representations that accompany those structures (the elastic ether, space as a receptacle for objects independent of those objects) are often replaced. There is a turnover in representations but a steady refinement of mathematical structure. So the terms 'unrepresentative realism' and 'structural realism' both have their point.

An important feature of progress in physics is the extent to which a theory can explain the degree of success enjoyed by the one it supersedes that goes beyond merely being able to reproduce its predictive success. Fresnel's theory of light was successful because, under a wide variety of circumstances, light does indeed have wave-like properties, a fact reinforced, not refuted, by contemporary theory. Similarly, from the point of view of relativity theory it can be appreciated why, under a wide variety of circumstances, involving masses that are not too great moving at speeds not too close to the velocity of light, treating space as a receptacle independent of time and of the objects in it, is an assumption that will not lead us far wrong. Any account of progress in physics needs to be able to accommodate such general features. What one calls the position that can accomplish this is of much less importance.

Further reading

This discussion has relied heavily on John Worrall's 1982 and 1989b texts. A collection of papers on scientific realism is Leplin (1984). Popper's defence of realism over instrumentalism is in his 1969 text (chapter 3) and his 1983 text. Classic defences of anti-realism are Duhem's 1962 and 1969 texts and Poincaré (1952), and a modern version is van Fraassen (1980). Scientific realism is defended in Psillos (1999), structural realism in Ladyman and Ross (2007) and anti-realism in Stanford (2006).

CHAPTER 16

Epilogue to the third edition

In this concluding section I offer some reflections on what has been achieved in the foregoing chapters. I raise three interrelated questions or problems that have concerned me during the writing of this book, and continue to do so.

1. Have I answered the question that forms the title of this book? What is this thing called science?
2. What is the relation between the historical examples given in the book and the philosophical thesis defended? Do the examples constitute evidence for my case, or are they simply illustrations?
3. How do the general claims made about science by the Bayesians and new experimentalists, discussed in chapters 12 and 13, relate to the case against method made in chapter 11? Isn't it the case that if there is no general account of science then all further discussion of the issue is redundant?

My response is as follows: I reaffirm that there is no general account of science and scientific method to be had that applies to all sciences at all historical stages in their development. Certainly philosophy does not have the resources to provide such an account. There is a sense in which the question that forms the title of this book is misguided. Nevertheless, a characterisation of the various sciences at various stages is a meaningful and important task. In this book I have attempted to accomplish that task for the physical sciences from the time of the scientific revolution in the seventeenth century until the present time (although I have refrained from tackling the question of the extent to which modern innovations such as quantum mechanics and quantum field theory involve characteristics that are qualitatively new). This task involves displaying the nature of the physical sciences mainly by means of historical examples of the appropriate kind. The historical examples therefore constitute an important part of the case rather than being mere illustrations of it.

Although the account of the physical sciences presented falls far short of providing a universal definition of science, it is far from useless when it comes to debates about what is or is not to count as science, exemplified, for example, in disputes about the status of 'creation science'. I presume that the main aim of those who defend creation science under that name is to imply that it has a character similar to that of acknowledged sciences such as physics. The position defended in this book enables that claim to be appraised. Having displayed what kind of knowledge claims are sought in physics, what kinds of methods are available for establishing them and what kind of success has been achieved, we have what we need as the basis for a comparison with creation science. Once the similarities and differences between the disciplines have been displayed, we have all that we need for a judicious appraisal of them, and will be in a position to appreciate whether anything can be legitimately read into the naming of creation science as a science. No universal account of science is necessary.

In the paragraph before last I indicated that my portrayal of the physical sciences is to be defended by reference to 'historical examples of the appropriate kind'. Some elaboration is called for here. Examples of the appropriate kind are concerned with the way in which the physical sciences function as *knowledge*. They are concerned with the kinds of claims made about the world in the physical sciences and the kinds of ways those claims are brought to bear on the world and tested against it. They are concerned with what philosophers call the *epistemology* of science. Philosophy of science is conducted by way of historical examples that display and clarify the epistemological function of science. The kind of history of science involved is a selective kind of history, and certainly not the only kind of history of science that is possible or important. The production of scientific knowledge always takes place in a social context in which that aim is interrelated with other practices with different aims, such as those involving the personal or professional aims of scientists, the economic aims of funding agencies, the ideological interests of religious or political groups of various kinds and so on. A history that explores these connections is both legitimate and important, but, I claim, beside the point as far as the project of this book is concerned. There is a range of kinds of 'social studies of science' currently in vogue that imply that an epistemological study of the kind I have conducted in this book cannot be achieved without due attention to the full range of senses in which science is social. In this book I have not faced the challenge posed by these schools of thought head-on. I have been content to show that what they say cannot be done can indeed be done simply by doing it. My attempt to square accounts with contemporary social studies of science appears in my *Science and Its Fabrication* (1990), a book in which I hope I make it clear that I regard a study of the social and political aspects of science as of great importance. The point at issue is the epistemological relevance of such studies.

Let me now turn to the question of the status of Bayesianism and the new experimentalism in the light of my denial of universal method. Bayesianism appears as an attempt to give an account of scientific reasoning in general, as is clearly signalled by the title of Howson and Urbach's 1993 text. However, this impression does not bear analysis. Even if we accept unquestioningly the Bayesian machinery, what that machinery gives us is a general way of adjusting the probability to be ascribed to beliefs in the light of new evidence. It does not single out scientific reasoning and distinguish it from other areas. Indeed, the most useful applications of Bayesianism are in gambling rather than science. Consequently, if Bayesianism is to tell us something distinctive about science in particular, then it will need to be augmented by some account of the kinds of beliefs and the evidence bearing upon them that occur in the sciences. I suggest that this can only be done by a careful look at the sciences themselves. What is more, I suggest that when that is done differences in the various sciences, and even qualitative changes within the methods of a single science, will emerge. That is, even if the Bayesian approach is the correct one, it does not stand as a threat to the denial of universal method, and is in need of the kind of epistemological history of science that I advocate.

The new experimentalists have certainly revealed some important features of experiment and its achievements within the physical and biological sciences. However, the account of science that this yields cannot be taken as providing *the* universal account of science. By way of examples, the new experimentalists have demonstrated the capabilities and achievements of experiment in the natural sciences during the last three hundred years and Deborah Mayo has provided a formal underpinning for much experimental reasoning by appeal to error theory and statistics. This does not amount to a universal account of science for two reasons. First, the emphasis on experimental manipulation involved in the new experimentalism renders that account largely irrelevant for an understanding of disciplines, especially in the social and historical sciences, where experimental manipulation is impossible or inappropriate. This conclusion could conceivably be avoided by identifying science with experimental science, but this would hardly serve to appease those who wish to call themselves political scientists or social scientists, for example. Second, as was argued in chapter 13, the new experimentalist account is incomplete insofar as it does not include an adequate account of the various crucial roles played by theory in science. The problem is very evident, I suggest, in Peter Galison's 1997 text in which he gives a descriptively rich account of progress in twentieth-century microparticle physics by focusing on the particle detectors and counters, their capabilities and their evolution. What is left unclear in the book is the relation between the experimental detection of particles and the high-level theory, involving symmetry and conservation principles, by means of which the particles are understood and classified. At the time of writing this epilogue, I regard it as an outstanding and pressing problem in the philosophy of the natural sciences to augment the insights of the new experimentalists with a correspondingly updated account of the role or roles of theory in the experimental sciences, substantiated by detailed case studies.

The following historical reflection illustrates the difficulty of extracting some universal characterisation of or prescription for science from the work of the new experimentalists, and illustrates the kind of study I have in mind for clarifying the nature of the relationships between theory and experiment. The idea that one should attempt to understand the world by experimentally manipulating it was by no means novel at the time of the scientific revolution. Alchemy, understood broadly as the precursor to modern chemistry involving the purposeful transformation of matter rather than narrowly as the attempt to transmute metals into gold, dates back to antiquity and flourished in the medieval period. The practice was not particularly successful. That lack of success cannot be simply attributed to lack of guidance by theory. A range of atomistic and other matter theory informed the work of the alchemists. If one is inclined to ignore theory and look simply to experimental practice, then significant progress can be discerned in the craft traditions of the metallurgists and drug manufacturers of the sixteenth and seventeenth centuries. However, the knowledge involved can be seen as qualitatively different from the chemistry that was to emerge in the late seventeenth and eighteenth centuries. The latter did involve 'theory', but very low-level theory far removed from atomism. What was needed, and what was supplied early in the eighteenth century, was a notion of chemical combination and recombination of substances, involving the idea that substances, when combined, continue to exist in the resulting compound and are there to be extracted again by means of appropriate manipulations. The classification of substances into acids and alkalis, and the salts produced by the neutralisation of one by the other, offered a way of organising research in a way that made progress possible without the need for some atomistic or other matter theory. It was well into the nineteenth century before the time was ripe for such speculations to be linked with experiment. So the question of the role of experiment in science and its relation to theory is a complex and historically relative one even if we restrict the discussion to chemistry.

I conclude with some remarks about the relationship between the views on science explored in this book and the work of scientists. Since I have denied that there is a universal account of science available to philosophers and capable of providing standards for judging science, and since I have argued that an adequate account of various sciences is only to be had by way of a close look at the sciences themselves, it might be concluded that the views of philosophers of science are redundant and that only those of scientists themselves are of consequence. It might be thought, that is, that insofar as I have successfully made my case, I have done myself out of a job. This conclusion (fortunately for me) is unwarranted. Although it is true that scientists themselves are the practitioners best able to conduct science and are not in need of advice from philosophers, scientists are not particularly adept at taking a step back from their work and describing and characterising the nature of that work. Scientists are typically good at making scientific progress, but not particularly good at articulating what that progress consists of. This is the reason that scientists are not particularly well equipped to engage in debates about the nature and status of science, and do not typically do a good job when it comes to controversies about the nature and status of science such as are involved, for example, in the evaluation of creation science. This book is not intended to be a contribution to science, not even the physical sciences on which I have focused. Rather, largely by means of historical examples, I have tried to clarify what kind of things the physical sciences are or have been.

Further reading

For an account of alchemy in the medieval period, and the various atomistic theories involved, see Newman (1994). The case for interpreting alchemy as chemistry rather than more narrowly, and an account of the invention of the narrow interpretation of 'alchemy' at the turn of the seventeenth century, can be found in Newman and Principe (1998). For an account of the introduction of an account of chemical combination capable of sustaining the new science of chemistry in the eighteenth century, see Klein (1995) and Klein (1996).

CHAPTER 17

Postscript

Introduction

As is implied by its title, my book *The Scientist's Atom and the Philosopher's Stone: How Science Succeeded and Philosophy Failed to Gain Knowledge of Atoms* is an attempt to distinguish between scientific knowledge of and philosophical speculations about atoms. I aim to pinpoint features of the knowledge of atoms that had emerged by the early twentieth century that warrants it being referred to as science in a way that was not the case for previous versions of atomism. We get a firm handle on what this thing called science is by appreciating the difference between the status of atomism in physical science at the turn of the nineteenth century and atomism as it occurs in the writings of the likes of Democritus or the seventeenth-century mechanical philosophers.

A rough characterisation of my account of the difference between scientific and philosophical claims concerning atoms is that the former are empirically confirmed in the way that the latter are not. The idea that what is distinctive about scientific knowledge lies in the extent to which it is confirmed by appeal to experience is hardly novel. It is the commonsense idea that formed the starting point for *What is this thing called Science?* and which, in the subsequent chapters, was subject to critical scrutiny. In the Introduction I noted that 'the idea that the distinctive feature of scientific knowledge is that it is derived from the facts of experience can only be sanctioned in a carefully and highly qualified form, if it is to be sanctioned at all' (p. xx). I have become less guarded and perhaps more conservative in my old age. I believe that once we have the appropriate notions of confirmation and empirical evidence, the distinguishing feature of science lies precisely in the sense in which it is empirically supported.

The idea that scientific knowledge is distinctive by virtue of the way in which it is confirmed by empirical evidence requires an appropriately demanding notion of confirmation and of what counts as evidence. Claims are borne out to the extent that they are tested against, rather than merely accommodated to, the evidence, whilst the relevant evidence is typically that arising from exacting experimental interventions. I do not have much to add to my discussion of the nature and status of experimental results to what can be found in the foregoing chapters, especially chapter 13. But I do wish to elaborate on and add to the discussion of confirmation.

A theme running through *What is this thing called Science?* and, indeed, through contemporary philosophy of science generally, is the fallibility of scientific knowledge. General claims made within science are fallible because of the limited scope and fallibility of the evidence on which they are based. There is also the historical point that theories once regarded as well confirmed have been found wanting and superseded. There is a tension between this fallibility and the idea that scientific theories are special because they are well confirmed. Part of the response to the difficulty is to insist on an appropriately strong form of confirmation. Aristotle's theory involving the four elements has been rejected, sure enough, but then it was never confirmed in the appropriately strong sense. On the other hand, for instance, Newton's mechanics was confirmed in a very strong sense, but this did not prevent it breaking down for masses that are very large or moving very fast. There is a sense in which Newton's theory has been falsified but not to the extent that it has been discarded. It is a limiting case of relativity theory and is still used to predict the paths of spacecraft and artificial satellites. A characteristic of science is that well-confirmed theories live on as limiting cases of their successors. There is a sense in which theories are fallible however well confirmed, but it is a highly qualified sense.

The idea that scientific theories are such by virtue of being well confirmed but are nevertheless liable to be superseded has consequences for the debate between realists and anti-realists. If realists are taken to be arguing that science progresses towards a more and more accurate picture of what the world is ultimately like, then the extent to which theories, however well confirmed, are liable to be superseded by theories involving radically different pictures poses a serious difficulty. If we consider the question of whether the physical world is ultimately continuous or discontinuous, for example, it is difficult to understand the progress of physical science as moving us closer to an answer to that question. The case for an atomic structure of matter was compelling by the early twentieth century, but it was not long before they were being described by continuous wave functions stretching to infinity. What is more, continuous electromagnetic fields became as basic to physics as atoms, with both exhibiting wave-like as well as particle-like characteristics. Who knows what deep structure might be attributed to the electromagnetic field or the electron in the light of revelations made possible by the next generation of particle accelerators? Such reflections pose problems for strong versions of scientific realism. On the other hand, the evidence for the existence of electrons is so strong that it is silly to imagine that future science will show that they do not exist, although it might well reveal some surprising things about them. Later in this Postscript I revisit the realism/anti-realism debate that I discussed in chapter 15 to try to do a better job of pinpointing the sense in which science warrants a realist interpretation.

Confirmation by arguments from coincidence

However well confirmed they might be, the claims of science are liable to be found wanting when they are pushed to new levels of accuracy or when they are applied in areas where they have never been tested before. How could it be otherwise? But this sense in which science is fallible does not alter the fact that laws and theories can be confirmed to such a high degree that they cannot possibly be completely on the wrong track. If claims are genuinely confirmed by a range of independent evidence then this would be an inexplicable coincidence unless they were good approximations to the truth.

Arguments from coincidence work only to the extent that they are borne out by evidence in a sufficiently strong sense. There is a range of factors of relevance to their strength.

First, observational and experimental evidence itself must have survived a range of stringent, objective tests along the lines considered in chapters 2, 3 and 13.

Second, if claims are to be borne out by evidence then they must be genuinely tested against rather than accommodated to that evidence. It is no coincidence that Ptolemy's astronomy can be rendered compatible with observations of planetary positions if there is no limit to the degree to which epicycles can be added to ensure a fit with those observations.

A third factor is involved here. It arises from the fact that laws and theories need to be augmented by additional assumptions, auxiliaries for short, before they can be put to the test. A danger that must be guarded against is that success or failure of tests is due to inadequacies in the auxiliaries rather than in the claim under test. Newton's particle theory of light yields the law of refraction if it is assumed that the particles of light are attracted as they pass from an optically less dense to a more dense medium, moving at a greater speed in the latter as a result. But both of the latter assumptions are false and support for the theory illusory. Conversely, falsifications can be mistaken too. Given the estimates of the distance of the stars common at the time it was put forward, the Copernican theory clashed with the fact that parallax of the stars was not observable. But it was the assumption about the distance of the stars that was at fault, not the theory. The antidote to these problems is to insist that auxiliaries, where they are necessary, are subject to tests independent of the theory that is being argued for. To argue that the departure of the orbit of Uranus from that predicted by Newtonian astronomy must be due to the attraction of some unidentified body is to merely accommodate the theory to the evidence. But the situation was different once the planet Neptune was observed and its approximate orbit determined. The fact that the observed orbit of Uranus is in line with what is predicted by Newtonian theory once the attraction between Uranus and Neptune is taken into account constitutes a genuine confirmation rather than falsification of the theory.

A fourth factor, that is implicit in the discussion of chapter 13 but not sufficiently highlighted, concerns what Deborah Mayo (1996) has referred to as the partitioning of theories. Theories are not confirmed by evidence if they generalise to a greater degree than is warranted by the evidence. Newtonian theory, complete with the assumption of an absolute space, was borne out by a wide range of predictions of terrestrial and astronomical phenomena. But those predictions follow just the same if the assumption of an absolute space is dropped. The predictions of Newtonian theory require specification only of motions of bodies relative to each other. It is possible to partition Newton's laws from the assumption of absolute space. Once this is done, it can be appreciated that the latter assumption is redundant and so not tested against the evidence. It is no coincidence that Newton's theory complete with the assumption of absolute space can predict a wide array of evidence because that evidence is explained by the theory minus absolute space. The ways in which a theory can be partitioned are not necessarily transparent. It has become possible within the general theory of relativity to distinguish the assumption that space is curved from stronger assumptions about the degree and cause of the curvature in a way that was not fully appreciated in Einstein's time. Once this is done, it can be appreciated that some crucial predictions of general relativity theory require only curved space-time and not the more specific assumptions implicit, for example, in Einstein's own version of the theory. The experimentally detectable red shift is an example. From the point of view that I, following Mayo, am urging, Einstein's version of general relativity was not confirmed by the red shift even though it predicted it!

A fifth point concerns the range and variety of evidence for a theory. The wider the range of phenomena that confirm a theory in a way that conforms to the strictures outlined in the previous paragraphs, the stronger the degree of confirmation of the theory as a whole. It would indeed be a coincidence that a theory could survive a range of qualitatively different genuine tests if it bore no resemblance to the truth. Newton's theory explained details of planetary orbits to an ever increasing accuracy, the precession of the equinoxes, the tides, billiard-ball collisions, the decrease of gravity with height, and the return of Halley's comet. Each instance of confirmation met the standards outlined above. How could Newton's theory have accomplished this if it were false? Well, there is a sense in which it is false as we now know. It breaks down in the case of relativistic and quantum effects. But it is not radically false insofar as it truly describes to a good degree of approximation a wide range of phenomena. What is more, the fact that it is able to do this can be largely accounted for by the theories of relativity and quantum mechanics that have superseded it.

The idea that theories are more strongly confirmed the wider the range of evidence makes the strength of confirmation a matter of degree. The wider the range of phenomena by which a theory is confirmed the greater the coincidence that would be involved were the theory false. Ptolemy's theory of astronomy was genuinely confirmed to some degree. Epicycles were introduced to account for retrograde motion. But once this is done, then the theory predicts that planets will appear brightest when they are retrogressing because that is when they are nearest the earth. The confirmation of this prediction constitutes genuine evidence for Ptolemy's theory because there is independent evidence for the epicycles involved in it. But Ptolemy's theory is false nevertheless. I am not suggesting for a moment that degrees of confirmation can be quantified, but I hope my examples make it clear that there is a sense in which Newton's theory was much more strongly confirmed by the evidence I listed in the previous paragraph than Ptolemy's was by the correlation between retrograde motion and brightness.

The fallibility of science is often stressed by listing theories, such as Aristotle's theory of the elements, Newton's particle theory of light, the caloric theory of heat and so on that have since been rejected. The implication is that just as theories in the past have been rejected, in spite of the evidence in their favour, so our current theories will come to be rejected in the future. I hold that this overstates the case. Many theories of the past that have been rejected were not confirmed, or were only very weakly confirmed, by evidence once we have sufficiently high standards about what counts as confirmation. In those cases where they were well confirmed, theories are not totally rejected but live on as approximations to or limiting cases of their successors.

My recent study of atomism in Chalmers (2009) involves a detailed comparison of philosophical theories of atoms that originated in Ancient Greece and scientific theories of atoms that were to become part of physical science. I argue that the difference lies precisely in the fact that the latter were empirically confirmed while the former were not. In the next three sections I draw on this study to further illustrate and elaborate on the position on confirmation that I have outlined above.

Philosophical versus scientific knowledge of atoms

There are two crucial differences between the form of atomism that emerged in Ancient Greece and the atomism that has become basic to contemporary science. Philosophical atomism sought ultimate explanations of physical reality whereas contemporary science does not. Connected with this is the point that the latter is confirmed by empirical evidence whereas the former was at best merely accommodated to the evidence.

The pre-Socratic philosophers, those individuals who initiated an extremely abstract, rational mode of thinking about the nature of the universe in the sixth and fifth centuries BC before the time of Socrates, saw change as a problem that needed an answer. The problem was posed acutely by Parmenides, who argued the extreme and startling thesis that change is impossible and so must be an illusion. He began with the plausible assumption that something cannot come from nothing. But any change involves something coming into being that did not previously exist. That is, it involves something coming from nothing. Parmenides also held that the void, empty space, cannot exist because void is nothing, and 'nothing' cannot exist. He concluded that, contrary to how it seems, the universe is a homogeneous, unchanging sphere. Atomism was developed by Leucippus and Democritus in the fifth century BC as a response to Parmenides. Their atoms were minute parcels of being with an unchanging shape and size separated from each other by void. The latter was understood to exist in spite of the fact that it is, in a sense, nothing because it was regarded as the absence of being, that is, the absence of atoms. This picture of the ultimate structure of the world was sufficient for the atomists to render change possible. It comes about via the motion and rearrangement of atoms that are in themselves changeless.

I suspect the ancient atomists would have regarded the

suggestion that their theory should be defended empirically somewhat crass. Atomism was put forward as the ultimate explanation of change in general. For change to be intelligible there must be something that persists that allows us to say that whatever has changed retains its identity insofar as *it* has changed. We want to be able to say that the green leaf turned brown as well as the green leaf was replaced by a brown one. When we push our account of change far enough we reach the ultimate account that must involve an ultimate reality that persists through change and is the underlying substratum of all change. Observation of change will not reveal the reality that lies behind change. The appearances, accessible to the senses, will never yield knowledge of the reality that lies behind the appearances.

The atomists like Democritus did offer empirical support for their thesis in a weak sense. They did attempt to offer a plausible story of how the universe as we know it is compatible with the idea that it consists of nothing other than unchanging atoms in the void. So, for instance, they offered an account of how earth came to be as the result of the chance collisions of atoms, of how gravity results from atomic collisions, how perception arises as the result of atoms emanating from a perceived object reaching and interacting with our senses and so on. It is doubtful whether the proposed mechanisms were indeed capable of explaining the phenomena they were invoked to explain. In any case there was no independent evidence for the existence and character of these mechanisms. Ancient atomism was at best accommodated to, rather than confirmed by, empirical evidence.

Something like Democritean atomism was revived by mechanical philosophers in the seventeenth century. An important difference was that they restricted it to the material world, exempting the non-material world of minds, souls, angels and God in a way that the ancients had not done. The replacement of an Aristotelian world view by the mechanical one is typically seen as part and parcel of the scientific revolution of the seventeenth century. In my new book I argue that this is misleading. I distinguish between the mechanical philosophy including atomistic versions of it defended, for example, by Robert Boyle and Isaac Newton and new sciences such as Boyle's pneumatics and

Newton's mechanics. The ground for my distinction is precisely that the former was merely accommodated to the phenomena whereas the latter were empirically confirmed. I will say a little about both Boyle and Newton in this respect.

Boyle himself came close to making the distinction I invoke between philosophy of the kind represented by his mechanical philosophy and experimental science. He distinguished between intermediate causes and explanations accessible to experiment and ultimate atomic explanations. In his pneumatics he claimed that a range of phenomena, such as the behaviour of barometers and syringes, is explained by appeal to the weight and elasticity of air. He supported his claims with a range of experiments, many of them involving the air pumps he devised for the purpose. He redesigned his experiments to eliminate objections of critics. By the time he had done, he had good grounds for claiming his pneumatics to be established as 'matters of fact'. The situation was different when it came to his mechanical philosophy. He claimed that the material world is composed of unchanging corpuscles of universal matter possessing only a definitive shape and size and some degree of motion or rest. To defend this view he proposed possible mechanisms that he hoped would be capable of reproducing common phenomena. For instance, he appealed to interlocking shapes of corpuscles to explain various chemical phenomena and motions of corpuscles to explain heat. It is doubtful whether his explanations were successful. In the cases of gravity and elasticity he openly admitted himself unable to devise plausible explanatory mechanisms. Even in those cases where adequate mechanisms could be devised, he had no independent way of testing for their existence. Unlike the case with his pneumatics, Boyle was, at best, only able to accommodate his atomism to evidence. It was not confirmed by the evidence.

Newton is justly famous for the mechanics contained in the *Principia*. He was able to defend his theory by reference to a range of empirical evidence in a way that conformed to, and represents one of the most impressive examples of, the exacting account of theory-confirmation I outlined in the second section. This is how I interpret Newton's claim that his mechanics was 'derived from the phenomena'. Newton contrasted this aspect of his mechanics with the mechanical world view presented in Descartes' *Principles of Philosophy*. In my language Newton argues that Descartes had merely accommodated his mechanical world view to the phenomena whereas he had confirmed his mechanics by reference to observable phenomena. However, in addition to devising his mechanics, Newton was also an atomist. As far as defending his atomism is concerned, he was only able to argue for the extent to which he could accommodate his atomism to the phenomena.

The atomism of Democritus, Boyle and Newton was put forward as an account of the ultimate nature of the material world in general and was supported to the extent that observable phenomena could be accommodated to it. The kind of atomism that entered science late in the nineteenth century differed from this both with regard to the kinds of claims it involved and with regard to the kinds of evidence offered in its support. The kinetic theory of gases that understood them to be composed of molecules in motion and atomism in chemistry, that understood chemical reactions as coming about via the combination of atoms and molecules, were not theories about the world in general but theories about specific phenomena involving the behaviour of gases and chemical combination. The adequacy of atomic theories did not depend on the explanations offered being ultimate explanations. It soon became necessary to attribute an inner structure to atoms. Finally, the experimental evidence that could be appealed to in support of atomism early in the twentieth century met the strong demands concerning confirmation that I summarised in the second section of this Postscript.

Strong evidence for the particulate structure of matter was provided in 1897 via J. J. Thomson's experiments on cathode rays. Cathode rays pass through evacuated glass tubes across which a high voltage is applied. They had been known for over four decades when Thomson began his studies. Taking advantage of improved vacuum technology. Thomson was able to deflect cathode rays in electric and magnetic fields. Using electromagnetic laws and Newton's laws of motion, both supported by much independent evidence, Thomson showed that these deflections provided evidence that the cathode rays were composed of fast-moving charged particles. He was able to measure the ratio of the charge to the mass of the particles. Other experiments enabled the charge of the particles to be estimated separately, so that the mass of the particles could be calculated. The measurements revealed that the mass of the particles was extremely minute, two thousand times smaller than the mass attributed to atoms on the basis of the kinetic theory of gases. Thomson's case, strong in itself, gained extra weight from the fact that similar conclusions were reached by other researchers from studies of phenomena other than cathode rays. Pieter Zeeman had arrived at the conclusion that atoms contain small negatively charged particles by studying the splitting of spectral lines by magnetic fields, and the values of the mass and charge he measured for the particles, now known as electrons, were in line with Thomson's.

The electron emerged not as the result of speculations about the structure of matter in general but out of attempts to understand specific phenomena such as cathode rays and spectral lines by experimenting on them. Electrons with a specified mass and negative charge were not contrived to accommodate the phenomena but were forced on experimenters by the phenomena. The minute size of the electron, and the fact that they are negatively charged only, were unanticipated surprises both to Zeeman and Thomson. The need to ascribe charge, as well as mass, to the particles arose as a result of the effect on them of electric and magnetic fields. Charge emerged as a concept in the course of developments in electromagnetic theory in the nineteenth century. It did not fit easily into the mechanical world view and many physicists, including James Clerk Maxwell, attempted to give a mechanical explanation of it, in terms of stresses and strains in a mechanical ether or some such thing. Thomson and Zeeman treated the charge of the electron as an unexplained primitive along with its mass. This move, which was a natural outcome of experimental investigations of electromagnetic phenomena, ran counter, and owed nothing, to speculations of the kind involved in the mechanical or any other philosophy. The trend continued once it was found necessary to attribute a half-integral spin to electrons, a quantum-mechanical notion with no classical correlate, and when it was realised that electrons obey quantum rather than classical statistics.

The atomic and sub-atomic physics that was evolving by the late nineteenth century involved claims and modes of argument that differed in kind from those involved in philosophical atomism such as that inspired by the mechanical philosophers. But that is not all. The emergence of the former owed next to nothing to the latter. Cathode rays were first produced by Julius Plücker in 1859 by taking advantage of some technical innovations. He was able to evacuate his discharge tubes to a degree not previously accomplished by employing the mercury diffusion pump devised by Johann Geissler, a technician in his own laboratory and he was able to generate high voltages using the induction coils devised by Heinrich Rühmkorff. It was not long before it was discovered that cathode rays are deflected by a magnet. Deflecting them in electric fields proved not so straightforward. When Hertz attempted it as late as 1884 he failed. Thomson's success in 1897 owed much to improved vacuum technology that had emerged as the result of attempts to increase the lifetime of electric light bulbs. There is a detailed story to be told that I only sketch here. But the sketch is enough to indicate that the path that led to Thomson's success does not bear much resemblance to the kinds of considerations involved in the articulation and defence of philosophical atomism.

Independent evidence and the 'theory-dependence of observation': Perrin's experiments on Brownian motion

Many of the points that I have been urging concerning the empirical confirmation of science can be well illustrated by a close look at the experiments on Brownian motion carried out by Jean Perrin from 1908 and the nature of the case for atoms that he was able to mount by appeal to them. Attention to the details shows the extent to which Perrin established the molecular constitution of gases by powerful arguments from coincidence, that his observations were not theory-dependent in a way that posed any kind of problem, and that the kinetic theory he defended was false in certain respects nevertheless.

In 1827 the British naturalist Robert Brown first observed through a microscope the agitated motion of minute particles suspended in a fluid. By late in the nineteenth century physicists suspected a link between what has become known as Brownian motion and the motions attributed to the molecules of fluids assumed in the kinetic theory. In support of this idea was the apparently chaotic nature of the motion, its persistence without an apparent cause and the extent to which it could be readily distinguished from the coordinated motions caused by external causes such as vibrations and local heating. In the way I analyse below, Perrin's experiments left no room for doubting that the motions of Brownian particles are indeed caused by collisions with molecules of the liquid in which they are suspended.

To capture the logic of Perrin's argument I will, in a more systematic way than Perrin in fact did, separate his experiments and the features of Brownian motion he was able to identify with them, on the one hand, and the way in which he could use his results to argue for the kinetic theory, on the other.

Perrin was able to closely observe Brownian particles using the ultra-microscope, invented in 1903.[1] By a number of independent methods Perrin was able to show that the motions were indeed random. An explanation was needed for the fact that the Brownian particles remained in suspension indefinitely and did not sink to the bottom of the container, in spite of the fact that the resin of which the particles were made was denser than the suspending liquid. An explanation was forthcoming based on the recognition that the number of suspended Brownian particles per unit volume decreased with height. Given this, the random motions of the particles give rise to an upward pressure, because the number of particles striking a horizontal layer in the suspending liquid from below would be slightly greater than that striking it from above. Equilibrium is attained when the effect of this pressure acting upwards balances the weight of the particles acting downwards. Perrin was able to show experimentally that the density distribution of particles did indeed remain constant indefinitely once thermal equilibrium had been reached, after three hours or so.

The pressure exerted by n particles of mass, m, striking a surface with average velocity, v, is a straightforward exercise in Newtonian mechanics. The change in momentum per unit time brought about by particles striking a surface could be calculated by Perrin employing statistical techniques common for almost half a century following James Clerk Maxwell's introduction of them in the context of the kinetic theory of gases. Setting the upwards force due to this pressure equal to the weight of the particles acting downwards yields the following equation:

$$W.\log(n_0/n) = 2\pi r^3\Delta.g.h$$

Here n_0/n represents the ratio of the number of particles per unit volume at two heights separated by a distance, h, Δ represents the excess of the density of the material of the Brownian particles over that of the liquid in which they are suspended, r is their radius, g is the acceleration due to gravity and W is the mean kinetic energy of the particles, that is, the mean value of $(1/2)mv^2$.

The equation shows n varying exponentially with height. Perrin was able to show experimentally that this was indeed the case. Further, since, in ways that are discussed below, Perrin was able to measure all the quantities in the above equation other than W, he could calculate the latter quantity. That is, Perrin was able to determine the mean value of the kinetic energy of Brownian particles. Series of experiments that varied the size and material of the particles and the nature of the suspending liquid revealed that the mean kinetic energy of the particles was independent of all these factors and depended only on the absolute temperature, T, at which the experiments were conducted.

The above discussion involves only what Perrin could establish on the basis of observations of Brownian motions assuming Newtonian mechanics. No appeal to the kinetic theory of gases is involved. But once it is assumed that the Brownian motion is caused by statistical fluctuations in the impacts experienced by the particles as they collide with molecules of the suspending liquid, as assumed in the kinetic theory, then Perrin's findings find a ready explanation. The broad features of the motion themselves require explanation. Here it needs to be realised that the Brownian particles rarely collide with each other. So what is the cause of the frequent changes in their direction of motion? Collisions of the particles with the random molecular motions of the suspending liquid supply the answer. It is a fundamental assumption of the kinetic theory that the temperature of a system is a measure of the mean kinetic energy of its underlying molecular motions. The fact that systems in thermal contact settle down to the same temperature means, from the point of view of the kinetic theory, that the mean kinetic energy of their molecular motions become equalised. Applying this assumption to Brownian motion, the mean kinetic energy of the Brownian particles will become equal to the mean kinetic energy of the molecules with which they collide once thermal equilibrium is reached. That is, the mean kinetic energy of the particles will be fixed once the temperature is fixed. We have an immediate explanation of Perrin's experimental finding that the mean kinetic energies he measured were independent of the size and material of the particles and the nature of the suspending fluid. Many quantitative features of Brownian motion as determined by Perrin have an immediate and straightforward explanation once we appeal to the kinetic theory but are mysterious otherwise.

The 'argument from coincidence' involved here by no means exhausts the case for molecules and their motions that Perrin was able to support by experiment. We have seen how he was able to measure the mean kinetic energy of Brownian particles at some absolute temperature, T. Fundamental to the kinetic theory is that the mean kinetic energy of the particles composing any system at temperature, T, will have the same mean kinetic energy. By measuring the mean kinetic energy of Brownian particles Perrin had not only measured the mean kinetic energy of the molecular motions responsible for their motion, he had measured the mean kinetic energy of the molecular motions of any system at temperature, T. In particular, he had in effect measured the mean kinetic energy of the molecules of a gas at temperature, T. This knowledge made it possible, in a straightforward way that I will not rehearse here, to calculate what is known as Avogadro's number, the number of molecules in a gram molecule of any gas. The fact that that value agreed well with values for the number arrived at by quite different methods further strengthened Perrin's 'argument from coincidence'.

There was yet more to come. The assumption that molecular motions are the cause of Brownian motion has implications for the mean displacement and mean rotation of the particles over time, as Einstein had shown in 1905. The kinetic theory implies that both of these quantities should be proportional to the square root of time elapsed. Perrin was able to show experimentally that this was indeed the case. What is more, measurements of the mean displacement and mean rotation gave Perrin two further ways of measuring the mean kinetic energy of the particles and hence of the molecules of a gas at a given temperature. This, in turn, gave him two further ways of calculating Avogadro's number. The results were in good agreement with the previous ones.

So far in my discussion I have simply assumed that Perrin could measure quantities such as the radius of Brownian particles and compare the number of particles per unit volume at different heights. Looking at some of the details of Perrin's measurements reveals important insights into the notions of theory-dependence of observation and the importance of independent tests. I take, as an example, the procedures by which Perrin established the radius of the Brownian particles, represented by 'r' in the above equation.

Reference to *the* radius of particles implies that they all have the same radius. It took Perrin up to three months to prepare emulsions containing particles of approximately equal size. He used a state-of-the-art centrifuge to do so. Measuring the size of particles was no straightforward matter. They were too small to be measured directly by a travelling microscope. In his first publication Perrin described how he estimated the size of particles by measuring the speed of fall of a cloud of particles through a liquid of known viscosity and calculating their radius using Stokes' law that relates the resisting force experienced by a sphere moving through a viscous medium to its velocity. Perrin was taken to task by critics, who pointed out that Stokes' law had not been experimentally confirmed for particles as small as Brownian particles, whilst the theoretical derivation of the law assumed that viscous forces vary continuously over the surface of the sphere, hardly an unproblematic assumption from the point of view of the kinetic theory! Perrin responded by devising two further methods of measuring the radius. Near the walls of the containing vessel the Brownian particles tended to coagulate in groups. Perrin looked amongst them for particles that were aligned. By measuring the length of, say, five aligned particles, which was possible with a travelling microscope, Perrin could estimate the radius of each. The fact that repetitions of the measurement for other aligned particles gave similar values for the radius gave Perrin direct evidence that the particles were indeed of uniform size. Perrin also estimated the size of particles by weighing a volume containing a countable number of them and calculating their radius using the density of the material of the particles, which he could also measure by several independent methods. With the exception of the ones that invoked Stokes' law, none of these measurements can be said to be theory-dependent in any useful sense. What is more, Perrin accepted values only when the results of several independent measurements agreed with each other. In similar kinds of ways which I will not document here, Perrin supported his measurements of the density of the material of which the Brownian particles were composed, the comparative numbers of particles per unit volume at various heights and the mean displacement and mean rotation of particles.

Perrin's experimental measurements were confirmed in as strong a sense as could reasonably be demanded. There are also some features of the derivation of the claims that Perrin put to the test that are of comparable importance for understanding the nature of his case for the kinetic theory. We have discussed how the addition of auxiliary assumptions to a theory introduces the possibility that imperfections in them can lead to false conclusions about the strengths and weaknesses of it. A striking feature of Perrin's argument is that it required only the basic assumptions of the kinetic theory without the need for auxiliaries. Basic to the kinetic theory was the randomness attributed to molecular motions and the equi-distribution of energy. Perrin gave novel support to the first by showing in a number of ways that the motion of the Brownian particles is indeed random. The equi-distribution of energy requires some discussion.

A basic assumption of the kinetic theory involves the identification of temperature with the mean kinetic energy of the molecules of a system. The molecules of all systems at the same temperature must possess the same mean kinetic energy. So, for example, if hydrogen gas is in thermal equilibrium with oxygen gas, the molecules of each gas must possess the same kinetic energy on average. Since oxygen molecules are heavier than hydrogen molecules they mostly, on average, move slower than the latter. Because, from the point of view of the kinetic theory, the only difference between molecules and Brownian particles with which they collide is one of scale, the equi-distribution of kinetic energy must apply to the particles too. This is the assumption, and the only assumption involving the kinetic theory, that was involved in Perrin's explanation of details of the density distribution of Brownian particles that we have discussed.

There is a more general sense in which the kinetic theory involves equi-distribution of energy. When molecules collide they lose or gain kinetic energy, not simply by transferring kinetic energy to or gaining it from each other but by gaining or losing rotational or vibrational energy. Nevertheless, if temperature is to remain constant the mean kinetic energy must remain unchanged by such exchanges of energy. On average, the kinetic energy lost to vibrational or rotational energy on some collisions must be balanced by kinetic energy gained in others. The result is an equi-distribution of energy between translational, rotational and vibrational modes. It is only on this assumption that a system can be said to have a definite temperature from the point of view of the kinetic theory. The equality of mean translational (kinetic) energy and rotational energy was involved in Perrin's investigation of the mean rotation of Brownian particles.

The extent to which Perrin's observations were strongly confirmed and the fact that his tests of the kinetic theory required only the general assumptions of that theory with no need for potentially problematic auxiliaries together account for the strength of his argument. Let me dramatise the situation in the case of Perrin's calculation of Avogadro's number from the density distribution of particles. Given the extent to which Perrin accepted only those measurements that were borne out by a number of independent methods, there is a strong sense in which Perrin could not choose the numbers he needed to feed into his equations. They were determined by the world, not him. The fact that the vertical distribution of Brownian particles should be exponential and that Avogadro's number could be calculated from it followed from the equi-distribution of kinetic energy basic to the kinetic theory and required no other assumptions. When Perrin measured the change in density distribution of Brownian particles over heights of around a tenth of a millimetre it was nature that delivered the resulting measurements. The result could conceivably have been that no variation in density was detected over such a small distance. Had this result eventuated Perrin would have calculated a value of zero for Avogadro's number. It might have been that given sufficient time the Brownian particles all settled to the bottom of the container. Had this eventuated Perrin would have derived a value of infinity for Avogadro's number. The range of possibilities could hardly have been larger. One can imagine why Perrin (1990, p. 104) greeted 'with the liveliest emotion' the fact that the number that his measurements actually led him to was around 7×10^{23}, close to previous estimates of Avogadro's number acquired by means not involving Brownian motion. And this was only part of the 'argument from coincidence' that Perrin was able to mount in favour of the kinetic theory.

Perrin's case for the kinetic theory could hardly have been stronger, and in the light of it one might be tempted to conclude that the insistence on the fallibility of scientific knowledge common in contemporary philosophy of science and given due credence in this book is misplaced. That would be a mistake. In fact, the kinetic theory supported by Perrin is false and known to be false. The equi-distribution of energy fundamental to the theory has straightforward consequences for the specific heats of gases that clash with observation. Equi-distribution breaks down for vibrational modes and, at low enough temperatures, for rotational modes as well, as Perrin himself became aware. It takes quantum mechanics to cope with the problem. After Perrin's experiments there could be little doubt about the molecular constitution of matter and the identification of molecular motions as the cause of Brownian motion. But this falls short of confirmation of the kinetic theory in an unqualified sense.

Partitioning of theories: atomism in nineteenth-century chemistry

Experimental support for theories is not always as compelling as that made by Perrin for the kinetic theory and, as we have noted, even when it is, we need to be circumspect about just what it is that has been confirmed. Perrin confirmed some aspects of the kinetic theory but not all of it. There are two issues here, both of which can be usefully illustrated by reference to John Dalton's introduction of atoms into chemistry. The first includes the recognition that the strength of arguments from coincidence is a matter of degree. The second is the need to partition theories where possible to help establish precisely which parts of them are confirmed by a particular experimental argument.

Dalton published his 'new system of chemical philosophy' in 1808. In it he proposed that each chemical element is made up of atoms and that chemical compounds are made up of 'compound atoms' consisting of atoms of the elements composing them. Atoms of an element are all alike as are compound atoms of a given compound. Dalton's theory can be put in its most favourable light by noting that it was borne out by three laws of proportion that it entailed. The law of constant proportions states that the relative proportions by weight of elements in their compounds are always the same. (The weight of oxygen to hydrogen in water is always eight to one.) The law of multiple proportions says that if two elements combine to form more than one compound then the weights of one element compared with a fixed weight of the other in each compound will be to each other as simple integral numbers. (The ratio of nitrogen per unit weight of oxygen in nitrous oxide, nitric oxide and nitrogen peroxide is as 4:2:1.) The law of reciprocal proportions states that if two elements A and B each combine with C to form compounds and if A and B combine to form a compound in the ratio x:y, then the relative weights of A and B that combine with a fixed weight of C will be as nx:my where n and m are small integral numbers. (Hydrogen and oxygen combine in the ratio 1:8 to form water, and both combine with nitrogen to form ammonia and nitric oxide. The weights of hydrogen and oxygen compared with a fixed weight of nitrogen in the latter compounds are as 3:16, which is 3x1 : 2x8, in accordance with the law.) It is not difficult to see how these laws can be seen as following straightforwardly from Dalton's assumption that least parts of compounds are made up of atoms of the elements possessing a definitive and invariable weight.

At the time he proposed it, Dalton's atomic theory received more opposition than support. There was a range of reasons why this was so. One reason lay in the fact that definitive atomic constitution of compounds could not be arrived at given the data available at Dalton's time, nor, indeed, for several decades thereafter. For instance, the experimentally determined fact that the weight of oxygen in water relative to hydrogen is eight is compatible with H_2O for water and an atomic weight of sixteen for oxygen relative to hydrogen, with HO for water and a relative atomic weight of eight for oxygen and a host of more complicated possibilities. A more fundamental reason is that chemistry at Dalton's time was not capable of being affected by the truth or otherwise of his atomic theory. Roughly speaking, chemists were interested in what combines with what and to what degree, and Dalton's atomism had nothing to say on that score. It should also be noted that my account of Dalton's theory is a cleaned-up version that abstracts the key chemical idea borne out by laws of proportion from much detail contained in Dalton's own version. For instance, Dalton assumed his atoms were surrounded by spherical clouds of caloric which he saw as having consequences for the specific heats of gases, he assumed that atoms of a given gas repel each other with a force inversely proportional to their separation and he assumed a simple relationship between atomic weights and the solubility of gases in liquids. These assumptions went nowhere. Dalton's atomic theory was supported to some degree by an argument from coincidence insofar as it could account for the three laws of proportion, but this fact was offset by the empirical difficulties associated with the extension of atomism beyond laws of proportion and the fact that the theory did not have the capacity to inform chemistry. The situation was to be transformed via the introduction of formulae into chemistry and their application to organic chemistry.

My use of formulae for water in the previous paragraph was not Dalton's own practice. He employed diagrams of arrangements of spheres to illustrate atomic combinations rather than formulae. The latter were first introduced into chemistry by the Swedish chemist Jacob Berzelius in 1813. It is important for my story about theory-confirmation to recognise that use of formulae and a commitment to atomism are not the same thing. As a matter of historical fact Berzelius introduced formulae as a device to capture what he saw as important about Dalton's innovation, namely, recognition of laws of proportion, *without a commitment to atomism* which he regarded as an interesting but problematic assumption. The symbols in a chemical formula, such as H_2O for water, *can* be interpreted as referring to atoms, and that is what is automatically assumed today. Given that interpretation, the formula for water together with measurements of combining weights imply that an atom of oxygen weighs sixteen times more than an atom of hydrogen. But there is another interpretation of the symbols in formulae that is more in keeping with what nineteenth-century chemists did in their laboratories. While an atom of hydrogen can be taken as the unit against which relative atomic weights are estimated, that is not necessary. Any portion of hydrogen whatsoever can be taken as supplying the reference weight. If that is done then the corresponding portion of oxygen will be one that is sixteen times heavier than the reference sample of hydrogen. The symbols in chemical formulae can be interpreted as referring to portions rather than atoms and this is sufficient to capture all the details about combining weights that the nineteenth-century chemists had access to and summarised with laws of proportion. Berzelian formulae were not much used in the couple of decades following their introduction, and this can be readily understood once it is recognised that they can be interpreted as a device for summarising facts about combining proportions that can be expressed in other ways. Dalton's opposition to their use perhaps stemmed from the fact that he recognised that they provided a way of representing his insights about combining proportions that avoided the commitment to atoms that he espoused.

The status of chemical formulae was transformed once they were employed to bring order into organic chemistry from the late 1820s. Their successful application in that area stemmed from the fact that they were used to express facts about chemical combination that went beyond laws of proportion. Prior to that success the choice between competing chemical formulae could plausibly be interpreted as conventional. On this view one can choose HO for water as long as one also chooses eight for the relative atomic weight of oxygen. There is no fact of the matter that makes this the wrong choice of formula and H_2O the right one. Developments in organic chemistry were to undermine the conventionalist position. By around 1860 organic chemists had arrived at formulae that they could argue to be the correct ones. The detailed story is a rich and complicated one. I illustrate with just one example. The simplest formula for acetic acid, given combining weights that can be measured and using modern values for relative atomic weights, is CH_2O. This formula cannot be used to reflect the fact that the hydrogen in acetic acid can be replaced in the laboratory by an equal volume of chlorine in four different ways, three of them yielding acids similar to acetic acid and a fourth yielding a salt. The situation can be handled by doubling the numbers in the formula and separating one of the hydrogens from the other three, resulting in $C_2H_3O_2H$. The three acids can now be seen as resulting from the replacement by chlorine of one, two or all three of the hydrogens grouped together and the salt resulting from replacement of the lone hydrogen, now identified as responsible for the acidic character of acetic acid, by chlorine. The demands put on formulae of the kind I have illustrated had, by around 1860, yielded a unique set of them up to the task. Chemists sought the right formulae, not just useful ones, and the conventionalist position was undermined.

It is important for my discussion of theory-confirmation to recognise that whilst these moves in organic chemistry settled the debate over whether formulae were conventional or not, it did not settle the issue of whether the symbols in the correct formulae represent atoms or portions. We can take chemistry of the late nineteenth century and partition it so that assumptions concerning atoms are separated from the remainder, which, as we have seen, does not require the removal of chemical formulae from that remainder. If I am right to claim that all the evidence that could be brought to bear on their theories by nineteenth-century chemists can be handled by those theories with atoms partitioned off, then it cannot be said that atomism was confirmed.

My claim that atomism in nineteenth-century chemistry was not well confirmed and was in principle redundant is controversial. It is not endorsed by the majority of contemporary scientists and philosophers. The attitude of nineteenth-century chemists themselves is more complicated. A position very similar to mine was argued by Pierre Duhem (2002) at the turn of the nineteenth century but his view was not the common one. He tends to be dismissed as a misguided positivist these days, but mistakenly as far as the situation in chemistry is concerned. Insofar as nineteenth-century chemists endorsed atomism, they were wary of identifying atoms with those assumed in the mechanical philosophy or the kinetic theory. They recognised that attributing properties to atoms sufficient to account for their role in chemistry was problematic and an issue that needed to be settled by research rather than philosophical decree. For instance, after around 1860 it became clear that atoms need to possess valency, and this property posed problems for, rather than followed from, mechanism and the physics of the time.

My, and Duhem's, reservations about the status of atomic theory in nineteenth-century chemistry notwithstanding, atomic chemistry was eventually confirmed experimentally, of course, and it did become possible and necessary to identify the atoms and molecules assumed in chemistry and physics. We have described in the previous section how Perrin's experiments on Brownian motion were sufficient to remove serious doubt about the existence of molecules, and the fact that the atomic and molecular weights that could be calculated via the kinetic theory coincided with those arrived at by chemists was just one of the increasing number of ways in which atomism in chemistry was confirmed in the twentieth century. But it is significant that it soon became necessary, in physics and chemistry, to attribute quantum mechanical properties to atoms that differed from anything anticipated by nineteenth-century chemists or physicists.

I have discussed the status of atomism in nineteenth-century chemistry mainly to illustrate the importance of the need to partition theories, where possible, in order to determine which of the partitioned parts can be said to be confirmed by specific experiments. But there is another general point about the nature and status of science that can usefully be raised here. In contemporary science it is taken for granted that the deep structure of the material world is very different from the world of everyday experience and features of it clash with many of our common intuitions. Sub-atomic particles possessing properties such as charm and parity that have no correlates in the macroscopic world of our experience are indispensable in fundamental particle physics while relativity theorists invoke multi-dimensional spaces and reject absolute simultaneity. The extent to which the deep structure of the world clashes with assumptions that serve us well in our everyday dealings with it was something that needed to be learnt. The seventeenth-century mechanical philosophers explicitly incorporated into their matter theories the assumption that the micro-world is like the macro-world in crucial respects so that knowledge of the latter can be projected onto the former. Their atoms were akin to and behaved like billiard balls. We now know how wrong they were. As far as knowledge of the material world is concerned, science is categorically distinct from the kind of metaphysics traditionally practised by philosophers and owes little to it. On this score, the intuitions of the positivists of the nineteenth and twentieth centuries were on the right track, even if they did make a mess of articulating them.

Realism versus anti-realism again

There have been developments in the philosophical literature in the debate between realists and anti-realists with respect to science that motivate me to look at the issue afresh. One of those developments has been further articulation, for example by Ladyman and Ross (2007), of what John Worrall (1989b) called 'structural realism'. The modification of realism involved in structural realism arises as a response to the strongest of the arguments against realism that invoke historically documented examples of the overthrow of previously successful theories. A key argument for realism appeals to the success of scientific theories involving unobservable entities such as atoms and electrons. The fact that such theories can successfully explain a range of phenomena and lead to the discovery of new phenomena would amount to a miracle or a remarkable coincidence if the entities, such as electrons and atoms, invoked by those theories did not exist and possess the properties attributed to them by the theories. The anti-realists respond by pointing to past theories, whose success included the novel explanations of phenomena, which posited entities that do not exist according to modern science. The success of the caloric theory and the theory that attributed light to waves in a material ether are the strongest examples of this kind. Structural realists admit the force of this argument and concede that science is not to be understood as providing or moving towards a correct description of the unobservable entities composing the world and the properties they possess. Rather, scientific theories are to be seen as identifying the structure of reality underlying observable phenomena. On this view, the wave equations involved in Fresnel's theory of light are good approximations to a true description of the structure of light even though there is no ether serving as the seat of the waves. This position is supported by the fact that Fresnel's equations follow as limiting cases of the electromagnetic theory of light. Fresnel was right insofar as the success of his theory stemmed from the attribution of a wave structure to light but wrong insofar as he implied the existence of a material ether.

For reasons that I explore below, I do not believe that structural realism is the correct response to the anti-realist case. In some respects its advocates overestimate the force of the anti-realist appeal to examples of theory change exhibited in the history of science. In another sense I believe that they fail to recognise a sense in which the anti-realist case is correct. My position draws on the strong sense of confirmation at work in science, the precise import of the correspondence theory of truth assumed by realists and the sense in which composites are no less real than the entities that compose them. These issues are discussed in turn in the following three subsections.

Strongly confirmed theories are never completely discarded

In the first section of this Postscript I summarised standards of confirmation that are very demanding. I contend that if we take these strong demands seriously, then we are in a good position to challenge the case against realism based on the history of science. The influential challenge to realists based on historical examples launched by Larry Laudan (1981) is a case in point. According to Laudan, past science is replete with theories, successful in their time, invoking such non-existent things as crystalline spheres, electrical fluids. light corpuscles and the like. He supports the so-called pessimistic induction. Just as few of the entities involved in past science refer to anything real from the standpoint of modern science, so few of the entities of contemporary science will be considered as referring to anything real from the point of view of future science. I respond as follows. Many of the theories invoked by Laudan never were confirmed, or were only very weakly confirmed, in the strong sense that I have argued has become distinctive of science. In those cases where confirmation was strong but the confirmed theories superseded nevertheless, tidied-up versions of the superseded theories lived on as limiting cases of their successors. The previously successful but now superseded theories did refer to something real insofar as the current theories of which they are limiting cases do.

The example that comes closest to supporting the anti-realist case involves the fate of the ether in nineteenth-century physics. As we have seen, it was mainly this example that led Worrall to formulate structural realism. There is no doubting that Fresnel's wave theory of light was strongly confirmed. However, my remarks about partitioning cast doubt on the claim that this confirmation extended to confirmation of the existence of a mechanical ether. The predictions that so strongly supported Fresnel's theory required only that light be a transverse wave, not that it be a wave in the ether. Evidence for a mechanical ether required evidence that went beyond establishing the wave character of light and it was in part the failure to produce such evidence that led to the ether being dropped from physics.

A natural objection to my position here is the observation that it involves hindsight. There is certainly a sense in which it does. It was a natural assumption to make in the mid-nineteenth century that a wave must be a wave in something. What is more, transverse waves required that that something be elastic. The conception of electromagnetic waves consisting of fluctuating fields in themselves that are not states of anything lay in the future. But this does not strike me as constituting a problem for realism. The realist idea that it is the aim of science to characterise the world at both the observational and unobservational level implies that claims about it can be false. Scientists have no direct access to the truth and all science will be forever subject to modification and improvement.

The arguments for molecules mounted by Perrin and for light

waves mounted by Fresnel were so strong that it is entirely implausible to suppose that some future science will show that there are no such things. This is not to deny that there remained much to be learnt about atoms at the time of Perrin and light at the time of Fresnel. An analogous range of points can be made with respect to the observable world. Our senses are imperfect and claims supported by them are fallible. Nevertheless, claims about observable objects can be substantiated by the extent to which they survive independent tests. If we suspect that the visible dagger is an illusion we can touch it. The conditions under which we are liable to have visual illusions are quite different from the conditions that lead to tactile illusions so it would be a remarkable coincidence if two sets of conditions should coincide to yield simultaneously the sighting and feel of a dagger when there is no dagger present. I suggest that the form of the argument from coincidence is what is important here, and is the same whether it is claims about the observable or the unobservable that is at issue. What is different is that in the case of the observable, independent tests are straightforwardly available through use of the senses whereas this is not the case for the unobservable. The comparison can be taken a stage further. It is quite possible to read into the evidence of the senses more than is warranted. It is not difficult to sympathise with those that supposed that fire is a very light substance that rises into the air by virtue of its lightness. 'Fire rises' is then a truth about this substance established by observation. It was eventually appreciated that there is no fire substance. But this does not mean that there is no such thing as flames. How could there not be given the range of evidence for them? I suggest that the situation involving the rejection of the ether and the retention of light waves is analogous, and, insofar as it is, the reality of the entities invoked by theoretical physics is on a par with that of observable objects.

Approximate truth is all we have

The following way of summarising a substantial part of the position I have outlined in the previous subsection has some appeal. Theories that are well confirmed are thereby shown to be approximately true and science progresses when they are replaced by theories that are better approximations to the truth. In the past I have refrained from formulating my position in this way because I have been aware of a challenging objection to it that stems from the problem of making precise sense of the notion of 'approximate truth'. Philosophers are able to give a precise account of truth as correspondence to the facts, so the objection goes, which is not matched by a corresponding account of what approximate truth amounts to. I am now inclined to meet this objection by challenging the premise on which it is based. While philosophers have been able to give a highly formal account of truth building on Tarski's insights they have not thereby altered the sense in which the correspondence theory of truth is a rough and ready conception. We only have a rough and ready notion of truth, or truth up to a point, but it is sufficient to make sense of commonsense and science and to support a realist interpretation of both areas of discourse.

I gave a brief account of Tarski's account of the correspondence theory of truth in chapter 15. Tarski showed that given the notion of the satisfaction of the predicates of a reasonably simple language the truth of propositions in that language can be explicated in a formal way that avoids difficulties typically associated with the notion of truth. As we have seen, his success stemmed from the distinction he made between the language whose truth-claims are being analysed, the object language, and the language being used to talk about the object language, the metalanguage. What Tarski did not supply was a formal account of how predicates in a language are satisfied by entities in the world. He did not give an account, for instance, of how it is that 'is white' or 'is a swan' can be satisfied by a particular instance of a swan. Insofar as the correspondence theory of truth is a theory of how sentences in a language can be true of the world, it presupposes an account of that relation which is essentially a commonsense account. An entity satisfies the predicate 'is white' if and only if it is white and satisfies 'is a swan' if it is a swan. Given this, Tarski is able to give a formal account of truth of generalisations involving whiteness and swans. 'All swans are white' is true if and only if all swans in the world are indeed white. The notion of correspondence between statements in a language or theory and the situations in the world that they describe is not explicated, but rather assumed, by Tarski.

Nor has this situation been changed by any subsequent improvements and extensions of Tarski's formalisations.

The notion of correspondence involved in the correspondence theory of truth exploited by realists is a non-technical notion having its origins in commonsense. As such it is a rough and ready notion. Treating it as such, we can straightforwardly appreciate how particular cases of correspondence can break down when claims are pushed beyond the realm in which satisfaction of the predicates involved make sense or when greater levels of precision are demanded. Perhaps it makes commonsense to say that an object falling straight down from a height moves faster than one rolling down a smooth inclined plane over the same height. But this needs considerable qualification from a more refined point of view. First, one needs to recognise that 'moves' does not distinguish between velocity and acceleration. It is true that the straight down motion has a greater acceleration than the inclined one. When it comes to velocity the situation is more complicated. The velocity of each motion changes from moment to moment, so if 'motion' means velocity the question concerning which moves faster is ill defined. If we interpret 'motion' as average velocity then one motion is not faster than the other since the average velocity is the same in the two cases. From the more refined point of view approximate truth can be recovered from the unrefined one. The claim that the straight down motion is the faster is true in the sense that it takes less time. What is more, it is happenings in the world that make it true. There is a straightforward sense in which the more refined description involving velocities and accelerations is a better approximation to the truth than the unrefined description involving motion, although there is a sense in which the unrefined claim can be said to be approximately true. We now know that if the refined notions are pushed into the realms of the very small or the very large and very fast they too prove to be in need of further refinement and turn out to be only approximations to the truth.

I have insisted that the notion of correspondence between linguistically formulated claims and the world they make claims about has its origins in commonsense. We do not first learn our everyday language and then learn how to apply it to the world. Learning how to manipulate and know the meaning of sentences involving 'swan' and applying talk of swans to instances of swans in the real world are parts of a single process. A correspondence theory of truth is implicit in our everyday usage of language. However, to recognise that correspondence has its origins in commonsense is not to say that it stays there. My example about speeds of falling is already suggestive of this. While rough estimates of times of fall are accessible to everyday observation, estimates of velocity are not. How could they be since they involve velocity at an instant, because velocity can vary from moment to moment. Estimates of the extent to which claims about the velocity of a moving object 'correspond to the facts' involve indirect measurements rather than direct observations. Nevertheless, as is exemplified in my discussion above (which is, incidentally, based on Galileo's criticism of commonsense notions and his own formulation of more precise ones), some continuity of reference is guaranteed insofar as the more refined notions arise as a response to problems with the crude ones. The new conception of motion fashioned by Galileo had a notion of correspondence built into it according to which it would seem as trite to say that 'object x has velocity v' is true if and only if object x has velocity v as to say that 'the swan is white' is true if and only if the swan is white.

As physics has probed deeper into the structure of matter, its concepts have departed more and more from commonsense ones and the ways in which its claims are tested empirically have become more and more indirect. So much so that the deep structure of matter as described by contemporary physics renders it inappropriate to talk of things possessing properties. At this level there is something to be said for the insistence of Ladyman and Ross (2007) that 'every thing must go'. But this need not be seen us undermining the notion of correspondence to the facts. It can still be meaningfully insisted that contemporary matter theory is true insofar as it corresponds to the facts however much the facts involved differ from those involving white swans or colliding billiard balls. Insofar as the abstract theories involved have arisen as an attempt to improve on earlier theories some degree of correspondence is built into the new theories. They are required to solve the problems with and emulate the success of their predecessor. The generation and interpretation of the new theoretical formulism are not separate steps. Assuming that they are unnecessarily makes attempts to achieve correspondence between the abstract theory and 'the facts' into a mystery.

Levels of reality

I resist the idea that when entities are explained by appeal to underlying structures and shown to be not quite what they were thought to be, they are thereby shown to be unreal. Liquids have a viscosity. Water flows through pipes more readily than oil does because it possesses a smaller viscosity than oil. The force with which a liquid resists the motion of a sphere through it is related to the viscosity of the liquid and the radius and velocity of the sphere by Stokes' law. These facts can be explained by the kinetic theory. They depend on the way that macroscopic motions in one region of a liquid are transported to other regions as a statistical result of the motions and collisions of the molecules that constitute the liquid. In the light of the explanation of viscosity by the kinetic theory it can be appreciated that claims about viscosity have their limits and break down when the density of the liquid or gas becomes very small. Indeed, at sufficiently low densities gases cease to have a viscosity because the statistical averages that give rise to it cease to be meaningful. It seems to me to be wrong-headed to conclude from this that only moving, colliding molecules, and not fluids with viscosity, exist. A raindrop runs down a eucalypt more readily than a globule of gum of the same size. Such happenings take place whether there is an observer present to witness them or not, and it is the happenings themselves that make statements describing them true or false. It has been acknowledged that at sufficiently low densities talk of viscosity breaks down. But this does not mean that sufficiently dense fluids have a viscosity that can be truly or falsely characterised. Crowds cease to be crowds when the individuals composing them disperse, but this does not mean that there is no such thing as a crowd which has a density and can surge.[2]

Anyone inclined to the view that the kinetic theory shows that there are really molecules in motion but not really liquids and gases with viscosity will soon find themselves denying that molecules are real. For the properties of molecules are explained by reference to their electronic structure. An electron in a molecule is analogous to a fly in a large railway station, so a molecule is mostly space and nothing like the miniature billiard ball that the kinetic theory treats it as. However, the electron itself may prove to have an inner structure that explains its properties. The view that in explaining entities by invoking more fundamental ones we thereby explain them away leads to ultimate realism. According to this view, the aim of science is to characterise the ultimate structure of reality. Such a characterisation will be true in an unqualified sense. It will correspond to the facts in an unqualified sense. Unlike talk of the viscosity of a gas or the trajectory of an electron, which can break down when pushed beyond its limits, the true description of ultimate reality will have no such limits.

Ultimate realism is open to serious objections. One of them is epistemological. How can we ever know that our most fundamental physics has reached the ultimate stage? If the history of physics is anything to go by, then successful theories get to be explained by deeper ones. What is more, the deeper theories invoke entities very unlike those they are able to explain. Given this, we might expect that the characterisation of reality involved in future fundamental physics will differ markedly from current fundamental physics. The ultimate characterisation of reality will not only be beyond the reach of science but it will also not be an end point which science can be sensibly said to approach.

A more serious difficulty with ultimate realism is implied by my characterisation of the correspondence theory of truth. Ultimate realism presupposes a theory that corresponds to the facts in an unqualified way. There are deep problems with what this could possibly amount to. Our theories are human constructions framed in languages, often mathematical languages, which are themselves human constructions. If there is an ultimate structure of reality it is not a human construction, so, presumably, the unqualified truths that describe it are not human constructions either. An ultimate realist who thinks that science progresses towards ultimate truth must envisage our humanly constructed theories converging on a view that is not a human construction at all. The notion of unqualified, objective correspondence to the facts assumed by the ultimate realist implies a relationship between statements about the world and the world itself that the correspondence theory of truth is unable to provide. We only have rough and ready correspondence or correspondence up to a point. But this rough and ready notion, implicit in commonsense and refined by science, is intelligible and is sufficient to make sense of science as a search for the truth and as progressing towards the truth. If anti-realism amounts to the claim that ultimate realism is incoherent or beyond the grasp of science then I am all for anti-realism.

Further reading

The distinction between scientific and philosophical atomism is analysed in historical detail in Chalmers (2009). Relevant debates concerning theory-confirmation are in Mayo and Spanos (2010). A recent articulation and defence of structural realism is in Ladyman and Ross (2007). Detailed analyses of the status and significance of Perrin's experiments on Brownian motion are in the interchange between van Fraassen (2009) and Chalmers (2011). Useful historical background can be found in Nye (1972).

Notes

Introduction

1. This list is from a survey by C. Trusedell cited in J. R. Ravetz (1971), p. 387n.

4. Deriving theories from the facts: induction

1. The quotation, by A. B. Wolfe, is cited in Hempel (1966, p. 11).

8. Theories as structures I: Kuhn's paradigms

1. Since first writing *The Structure of Scientific Revolutions* Kuhn has conceded that he originally used 'paradigm' in a number of different ways. In the Postscript accompanying the second edition he distinguishes two senses of the word, a general sense which he calls the 'disciplinary matrix', and a narrow sense of the term which he has replaced by 'exemplar'. I continue to use the word 'paradigm' in the general sense, to refer to what Kuhn now calls the disciplinary matrix.

10. Feyerabend's anarchistic theory of science

1. The quotation from Hume's 'Of the Original Contract' is in Barker (1976, p. 156). The specific views of Locke criticised in the passage can be found on pp. 70–2 of the same volume.

11. Methodical changes in method

1. My remarks in this paragraph should not be taken as implying that there is no room for a political and social analysis of science as it functions in society, as I try to make clear in *Science and Its Fabrication* (1990, chapter 8). Nor are my remarks intended to be dismissive of all that goes under the name of 'social studies of science' since much contemporary work has yielded valid insights into the nature of scientific work. They are directed only at those who consider themselves to have constructed sociological or other knowledge of such a high

status that from its point of view they can judge that scientific knowledge has no special status.

Appendix to chapter 13

1. I originally took my cases to be counter examples of Deborah Mayo's position too, but she convinced me otherwise in private correspondence.

17. Postscript

1. In the ultra microscope a specimen is observed by viewing the light scattered at right angles to the direction in which it is observed. In Perrin's experiments the light scattered by the molecules of the suspending fluid in that direction was negligible compared with that scattered by the Brownian particles, so the glare from the former was eliminated.
2. I agree with Alan Musgrave (1999, pp. 132–3) when he endorses the treatment of Eddington's two tables, the commonsense one and the molecular one, by Susan Stebbing (1937). As Musgrave says, when science explains the solidity of the table in terms of molecular interactions 'it does not show that the table is not solid as commonsense supposes'.

Bibliography

Ackermann, R. J. (1976). *The Philosophy of Karl Popper.* Amherst: University of Massachusetts Press.

Ackermann, R. (1989). 'The New Experimentalism', *British Journal for the Philosophy of Science,* 40, 185–90.

Anthony, H. D. (1948). *Science and Its Background.* London: Macmillan.

Armstrong, D. M. (1983). *What Is a Law of Nature?* Cambridge: Cambridge University Press.

Ayer, A. J. (1940). *The Foundations of Empirical Knowledge.* London: Macmillan.

Bamford, G. (1993). 'Popper's Explications of Ad Hocness: Circularity, Empirical Content and Scientific Practice', *British Journal for the Philosophy of Science,* 44, 335–55.

Barker, E. (1976). *Social Contract: Essays by Locke, Hume and Rousseau.* Oxford: Oxford University Press.

Barnes, B. (1982). *T. S. Kuhn and Social Science.* London: Macmillan.

Barnes, B., Bloor, D. and Henry, J. (1996). *Scientific Knowledge: A Sociological Analysis.* Chicago: University of Chicago Press.

Bhaskar, R. (1978). *A Realist Theory of Science.* Hassocks, Sussex: Harvester.

Bird, A. (2000). *Thomas Kuhn.* Chesham (England): Acumen.

Block, I. (1961). 'Truth and Error in Aristotle's Theory of Sense Perception', *Philosophical Quarterly,* 11, 1–9.

Bloor, D. (1971). 'Two Paradigms of Scientific Knowledge', *Science Studies,* 1, 101–15.

Boyd, R. (1984). 'The Current Status of Scientific Realism' in Leplin (1984) 41–82.

Brown, H. J. (1977). *Perception, Theory and Commitment: The New Philosophy of Science.* Chicago: University of Chicago Press.

Buchwald, J. (1989). *The Creation of Scientific Effects.* Chicago: University of Chicago Press.

Cartwright, N. (1983). *How the Laws of Physics Lie.* Oxford: Oxford University Press.

Cartwright, N. (1989). *Nature's Capacities and Their Measurement.* Oxford: Oxford University Press.

Chalmers, A. F. (1973). 'On Learning From Our Mistakes', *British Journal for the Philosophy of Science*, 24, 164–73.

Chalmers, A. F. (1984). 'A Non-Empiricist Account of Experiment', *Methodology and Science*, 17, 95–114.

Chalmers, A. F. (1985). 'Galileo's Telescopic Observations of Venus and Mars', *British Journal for The Philosophy of Science*, 36, 175–91.

Chalmers, A. F. (1986). 'The Galileo that Feyerabend Missed: An Improved Case Against Method' in J. A. Schuster and R. A. Yeo (eds), *The Politics and Rhetoric of Scientific Method.* Dordrecht: Reidel, 1–33.

Chalmers, A. F. (1990). *Science and Its Fabrication.* Milton Keynes: Open University Press.

Chalmers, A. F. (1993). 'The Lack of Excellency of Boyle's Mechanical Philosophy', *Studies in History and Philosophy of Science*, 24, 541–64.

Chalmers, A. F. (1995). 'Ultimate Explanation in Science', *Cogito,* 9, 141–5.

Chalmers, A. F. (1999). 'Making Sense of Laws of Physics' in H. Sankey (ed.), *Causation and Laws of Nature.* Dordrecht: Kluwer.

Chalmers, A. F. (2003). 'The Theory-dependence of the Use of Instruments in Science', *Philosophy of Science*, 70, 493–509.

Chalmers, A. F. (2009). *The Scientist's Atom and the Philosopher's Stone: How Science Succeeded and Philosophy Failed to Gain Knowledge of Atoms.* Dordrecht: Springer.

Chalmers, A. F. (2011). 'Drawing Philosophical Lessons from Perrin's Experiments on Brownian Motion: A Response to van Fraassen', *British Journal for the Philosophy of Science,* 62, 711–32.

Christie, M. (1994). 'Philosophers versus Chemists Concerning "Laws of Nature"', *Studies in History and Philosophy of Science,* 25, 613–29.

Clavelin, M. (1974). *The Natural Philosophy of Galileo.* Cambridge, Mass.: MIT Press.

Cohen, R. S., Feyerabend, P. K. and Wartofsky, M. W. (eds) (1976). *Essays in Memory of Imre Lakatos.* Dordrecht: Reidel.

Couvalis, G. (1989). *Feyerabend's Critique of Foundationalism*. Aldershot, Hampshire: Avebury.

Davies, J. J. (1968). *On the Scientific Method*. London: Longman.

Dorling, J. (1979). 'Bayesian Personalism and Duhem's Problem', *Studies in History and Philosophy of Science*, 10, 177–87.

Drake, S. (1957). *The Discoveries and Opinions of Galileo*. New York: Doubleday.

Drake, S. (1978). *Galileo at Work*. Chicago: Chicago University Press.

Duhem, P. (1962). *The Aim and Structure of Physical Theory*. New York: Atheneum.

Duhem, P. (1969). *To Save the Phenomena*. Chicago: University of Chicago Press.

Duhem, P. (2002). *Mixture and Chemical Combination and Related Essays*, transl. P. Needham. Dordrecht: Kluwer.

Duncan, M. M. (1976). *On the Revolutions of the Heavenly Spheres*. New York: Barnes and Noble.

Earman, J. (1992). *Bayes or Bust? A Critical Examination of Bayesian Confirmation Theory*. Cambridge, Mass., MIT Press.

Edge, D. O. and Mulkay, M. J. (1976). *Astronomy Transformed*. New York: Wiley Interscience.

Farrell, R. (2003). *Feyerabend and Scientific Values*. Dordrecht: Kluwer.

Feyerabend, P. K. (1970). 'Consolations for the Specialist' in Lakatos and Musgrave (1970), 195–230.

Feyerabend, P. K. (1975). *Against Method: Outline of an Anarchistic Theory of Knowledge*. London: New Left Books.

Feyerabend, P. K. (1976). 'On the Critique of Scientific Reason' in Howson (1976), 209–39.

Feyerabend, P. K. (1978). *Science in a Free Society*. London: New Left Books.

Feyerabend, P. K. (1981a). *Realism, Rationalism and Scientific Method. Philosophical Papers, Volume I*. Cambridge: Cambridge University Press.

Feyerabend, P. K. (1981b). *Problems of Empiricism. Philosophical Papers, Volume II*. Cambridge: Cambridge University Press.

Franklin, A. (1986). *The Neglect of Experiment*. Cambridge: Cambridge University Press.

Franklin, A. (1990). *Experiment, Right or Wrong.* Cambridge: Cambridge University Press.

Galileo (1957). 'The Starry Messenger' in S. Drake (1957).

Galileo (1967). *Dialogue Concerning the Two Chief World Systems,* transl. S. Drake. Berkeley, California: University of California Press.

Galileo (1974). *Two New Sciences,* transl. S. Drake. Madison: University of Wisconsin Press.

Galison, P. (1987). *How Experiments End.* Chicago: University of Chicago Press.

Galison, P. (1997). *Image and Logic: A Material Culture of Physics.* Chicago: University of Chicago Press.

Gaukroger, S. (1978). *Explanatory Structures.* Hassocks, Sussex: Harvester.

Geymonat, L. (1965). *Galileo Galilei.* New York: McGraw-Hill.

Glymour, C. (1980). *Theory and Evidence.* Princeton: Princeton University Press.

Goethe, J. W. (1970). *Theory of Colors,* transl. C. L. Eastlake. Cambridge, Mass.: MIT Press.

Gooding, D. (1990). *Experiment and the Making of Meaning: Human Agency in Scientific Observation and Experiment.* Dordrecht: Kluwer.

Hacking, I. (1983). *Representing and Intervening.* Cambridge: Cambridge University Press.

Hanfling, O. (1981). *Logical Positivism.* Oxford: Basil Blackwell.

Hanson, N. R. (1958). *Patterns of Discovery.* Cambridge: Cambridge University Press.

Hempel, C. G. (1966). *Philosophy of Natural Science.* Englewood Cliffs, N.J.: Prentice Hall.

Hertz, H. (1962). *Electric Waves.* New York: Dover.

Hirsch, P. B., Horne, R. W. and Whelan, M. J. (1956). 'Direct Observation of the Arrangements and Motions of Dislocations in Aluminium', *Philosophical Magazine,* 1, 677–84.

Hooke, R. (1665). *Micrographia.* London: Martyn and Allestry.

Horwich, P. (1982). *Probability and Evidence.* Cambridge: Cambridge University Press.

Howson, C. (ed.) (1976). *Method and Appraisal in the Physical Sciences.* Cambridge: Cambridge University Press.

Howson, C. and Urbach, P. (1993). *Scientific Reasoning: The Bayesian Approach*. La Salle, Illinois: Open Court.

Hoyningen-Huene, P. (1993). *Reconstructing Scientific Revolutions: Thomas S. Kuhn's Philosophy of Science*. Chicago: University of Chicago Press.

Hume, D. (1933). *Treatise on Human Nature*. London: Dent.

Klein, U. (1995). 'E. F. Geoffroy's Table of Different "Raports" Observed Between Different Chemical Substances', *Ambix*, 42, 79–100.

Klein, U. (1996). 'The Chemical Workshop Tradition and the Experimental Practice: Discontinuities Within Continuities', *Science in Context*, 9, 251–87.

Kuhn, T. (1959). *The Copernican Revolution*. New York: Random House.

Kuhn, T. (1970a). *The Structure of Scientific Revolutions*. Chicago: University of Chicago Press.

Kuhn, T. (1970b). 'Logic of Discovery or Psychology of Research' in Lakatos and Musgrave (1970), 1–20.

Kuhn, T. (1970c). 'Reflections on My Critics' in Lakatos and Musgrave (1970), 231–78.

Kuhn, T. (1977). *The Essential Tension: Selected Studies in Scientific Tradition and Change*. Chicago: University of Chicago Press.

Ladyman, J. and Ross, D. (2007). *Every Thing Must Go: Metaphysics Naturalized*. New York: Oxford University Press.

Lakatos, I. (1968). *The Problem of Inductive Logic*. Amsterdam: North Holland.

Lakatos, I. (1970). 'Falsification and the Methodology of Scientific Research Programmes' in Lakatos and Musgrave (1970), 91–196.

Lakatos, I. (1971). 'Replies to Critics' in R. Buck and R. S. Cohen (eds), *Boston Studies in the Philosophy of Science, Volume 8*. Dordrecht: Reidel.

Lakatos, I. (1976a). 'Newton's Effect on Scientific Standards' in Worrall and Currie (1978a), 193–222.

Lakatos, I. (1976b). *Proofs and Refutations*. Cambridge: Cambridge University Press.

Lakatos, I. (1978). 'History of Science and Its Rational Reconstruction' in Worrall and Currie (1978a), 102–38.

Lakatos, I. and Musgrave, A. (eds) (1970). *Criticism and the Growth of Knowledge*. Cambridge: Cambridge University Press.

Lakatos, I. and Zahar, E. (1975). 'Why Did Copernicus' Programme Supersede Ptolemy's' in R. Westman (ed.), *The Copernican Achievement* Berkeley, California: University of California Press.

Larvor, B. (1998). *Lakatos: An Introduction*. London: Routledge.

Laudan, L. (1977). *Progress and Its Problems: Towards a Theory of Scientific Growth*. Berkeley: University of California Press.

Laudan, L. (1981). 'A Confutation of Convergent Realism', *Philosophy of Science*, 48, 19–49.

Laudan, L. (1984). *Science and Values: The Aims of Science and Their Role in Scientific Debate*. Berkeley: University of California Press.

Leplin, J. (1984). *Scientific Realism*. Berkeley: University of California Press.

Locke, J. (1967). *An Essay Concerning Human Understanding*. London: Dent.

Maxwell, J. C. (1877). 'The Kinetic Theory of Gases', *Nature*, 16, 245–46.

Maxwell, J. C. (1965). 'Illustrations of the Dynamical Theory of Gases' in W. D. Niven (ed.), *The Scientific Papers of James Clerk Maxwell, 2 Volumes*. New York: Dover.

Mayo, D. (1996). *Error and the Growth of Experimental Knowledge*. Chicago: University of Chicago Press.

Mayo, D. and Spanos, A. (2010). *Error and Inference: Recent Exchanges on Experimental Reasoning, Reliability and the Rationality of Science*. New York: Cambridge University Press.

Menter, J. (1956). 'The Direct Study by Electron Microscopy of Crystal Lattices and Their Imperfections', *Proceedings of the Royal Society, A*, 236, 119–35.

Mill, J. S. (1975). *On Liberty*. New York: Norton.

Mulkay, M. (1979). *Science and the Sociology of Knowledge*. London: Allen & Unwin.

Musgrave, A. (1974a). 'The Objectivism of Popper's Epistemology' in Schilpp (1974), 560–96.

Musgrave, A. (1974b). 'Logical Versus Historical Theories of Confirmation', *British Journal for the Philosophy of Science*, 25, 1–23.

Musgrave, A. (1993). *Commonsense, Science and Scepticism: A Historical Introduction to the Theory of Knowledge*. Cambridge: Cambridge University Press.

Musgrave, A. (1999). *Essays on Realism and Rationality*. Amsterdam: Rodopi.

Nersessian, N. (1984). *Faraday to Einstein: Constructing Meaning in Scientific Theories*. Dordrecht: Kluwer.

Newman, W. R. (1994). *Gehennical Fire: The Lives of George Starkey, an American Alchemist in the Scientific Revolution*. Cambridge: Harvard University Press.

Newman, W. R. and Principe, L. M. (1998). 'Alchemy vs Chemistry: the Etymological Origins of a Historiographic Mistake', *Early Science and Medicine*, 3, 32–65.

Nye, M. J. (1972). *Molecular Reality: A Perspective on the Scientific Work of Jean Perrin*. London: MacDonald.

Nye, M. J. (1980). 'N-rays: An Episode in the History and Psychology of Science', *Historical Studies in the Physical Sciences*, 11, 125–56.

Oberheim, E. (2006). *Feyerabend's Philosophy*. Berlin: De Gruyter.

O'Hear, A. (1980). *Karl Popper*. London: Routledge and Kegan Paul.

Perrin, J. (1990). *Atoms*. Woodbridge, Connecticut: Ox Bow Press.

Poincaré, H. (1952). *Science and Hypotheses*. New York: Dover.

Polanyi, M. (1973). *Personal Knowledge*. London: Routledge and Kegan Paul.

Popper, K. R. (1969). *Conjectures and Refutations*. London, Routledge and Kegan Paul.

Popper, K. R. (1972). *The Logic of Scientific Discovery*. London: Hutchinson.

Popper, K. R. (1974). 'Normal Science and Its Dangers' in Lakatos and Musgrave (1974), 51–8.

Popper, K. R. (1979). *Objective Knowledge*. Oxford: Oxford University Press.

Popper, K. R. (1983). *Realism and the Aim of Science*. London: Hutchinson.

Price, D. J. de S.(1969). 'A Critical Re-estimation of the Mathematical Planetary Theory of Ptolemy' in M. Clagett (ed.),

Critical Problems in the History of Science. Madison: University of Wisconsin Press.

Psillos, S. (1999). *Scientific Realism: How Science Tracks Truth*. London: Routledge and Kegan Paul.

Psillos, S. (2002). *Causation and Explanation*. Chesham, England: Acumen.

Quine, W. V. O. (1961). 'Two Dogmas of Empiricism' in *From a Logical Point of View*. New York: Harper and Row.

Ravetz, J. R. (1971). *Scientific Knowledge and Its Social Problems*. Oxford: Oxford University Press.

Rosen, E. (1962). *Three Copernican Treatises*. New York: Dover.

Rosenkrantz, R. D. (1977). *Inference, Method and Decision: Towards a Bayesian Philosophy of Science*. Dordrecht: Reidel.

Rowbotham, F. J. (1918). *Story Lives of Great Scientists*. London: Wells, Gardner and Darton.

Russell, B. (1912). *Problems of Philosophy*. Oxford: Oxford University Press.

Salmon, W. (1966). *The Foundations of Scientific Inference*. Pittsburgh: University of Pittsburgh Press.

Schilpp, P. A. (ed.) (1974). *The Philosophy of Karl Popper*. La Salle, Illinois: Open Court.

Shapere, D. (1982). 'The Concept of Observation in Science and Philosophy', *Philosophy of Science*, 49, 485–525.

Stanford, P. K. (2006). *Exceeding Our Grasp*. New York: Oxford University Press

Stebbing, S. (1937). *Philosophy and the Physicists*. London: Methuen.

Stove, D. (1973). *Probability and Hume's Inductive Skepticism*. Oxford: Oxford University Press.

Thomason, N. (1994). 'The Power of ARCHED Hypotheses: Feyerabend's Galileo as a Closet Rationalist', *British Journal for the Philosophy of Science*, 45, 255–64.

Thomason, N. (1998). '1543 – The Year That Copernicus Didn't Predict the Phases of Venus' in A. Corones and G. Freeland (eds), *1543 and All That*. Dordrecht: Reidel.

Thurber, J. (1933). *My Life and Hard Times*. New York: Harper.

van Fraassen, Bas C. (1980). *The Scientific Image*. Oxford: Clarendon.

van Fraassen, Bas C. (1989). *Laws and Symmetry*. Oxford: Oxford University Press.

van Fraassen, Bas C. (2009). 'The Perils of Perrin, in the Hands of Philosophers', *Philosophical Studies*, 143, 5–24.

Woolgar, S. (1988). *Science: The Very Idea*. London: Tavistock.

Worrall, J. (1976). 'Thomas Young and the "Refutation" of Newtonian Optics: A Case Study in the Interaction of Philosophy of Science and History' in Howson (1976), 107–79.

Worrall, J. (1982). 'Scientific Realism and Scientific Change' in *Philosophical Quarterly*, 32, 201–31.

Worrall, J. (1985). 'Scientific Reasoning and Theory Confirmation' in J. Pitt (ed.), *Change and Progress in Modern Science*. Dordrecht: Reidel.

Worrall, J. (1988). 'The Value of a Fixed Methodology', *British Journal for the Philosophy of Science*, 39, 263–75.

Worrall, J. (1989a). 'Fresnel, Poisson and the White Spot: The Role of Successful Predictions in Theory Acceptance' in D. Gooding, S. Schaffer and T. Pinch (eds), *The Uses of Experiment: Studies of Experiment in Natural Science*. Cambridge: Cambridge University Press.

Worrall, J. (1989b). 'Structural Realism: The Best of Both Worlds?', *Dialectica*, 43, 99–124.

Worrall, J. and Currie, G. (eds) (1978a). *Imre Lakatos, Philosophical Papers, Volume I: The Methodology of Scientific Research Programmes*. Cambridge: Cambridge University Press.

Worrall, J. and Currie, G. (eds) (1978b). *Imre Lakatos, Philosophical Papers, Volume 2: Mathematics, Science and Epistemology*. Cambridge: Cambridge University Press.

Zahar, E. (1973). 'Why Did Einstein's Theory Supersede Lorentz's', *British Journal for the Philosophy of Science*, 24, 95–123 and 223–63.

Index of names